QUALITY CONTROL ENGINEERING AND MANUFACTURING

U.S. INDUSTRIAL DESIGN SECTOR

OCCUPATIONAL, BUSINESS AND INNOVATION PROFILES

QUALITY CONTROL ENGINEERING AND MANUFACTURING

Additional books in this series can be found on Nova's website under the Series tab.

Additional e-books in this series can be found on Nova's website under the e-book tab.

QUALITY CONTROL ENGINEERING AND MANUFACTURING

U.S. INDUSTRIAL DESIGN SECTOR

OCCUPATIONAL, BUSINESS AND INNOVATION PROFILES

LOUIS H. NIELSEN
EDITOR

New York

For permission to use material from this book please contact us:
Telephone 631-231-7269; Fax 631-231-8175
Web Site: http://www.novapublishers.com

NOTICE TO THE READER

LIBRARY OF CONGRESS CATALOGING-IN-PUBLICATION DATA

ISBN: 978-1-62948-371-9

Published by Nova Science Publishers, Inc. † New York

CONTENTS

PREFACE

This book brings together analytical perspectives regarding federal data on industrial design drawn from the U.S. Bureau of Labor Statistics (BLS). The BLS defines industrial designers as "those individuals that develop the concepts for manufactured products such as cars, home appliances, and toys." While this dataset enables a quantitative grasp of the industry, BLS' definition is arguably limited in scope. Today's industrial designers find themselves in a variety of roles and functions beyond the development of manufactured products.

Chapter 1 – Design is a field with a large and extensive presence in our nation's manufacturing and services industries, as documented by the national datasets that provide the basis for this report. Designers are prolifically inventing new products, processes, and systems that have a profound impact on our economy and civil society.

The National Endowment for the Arts (NEA) Design Program has been tracking numerous trends in the field of design, from the growing movement of design thinking to social impact design. Although this report brings together, for the first time, analytical perspectives regarding federal data on industrial design, it cannot be all-encompassing. This preface has benefited from conversations with some of the nation's leading designers, design curators, and design firms to convey information not captured by the report itself.

Chapter 2 – When I was in school back in the 1970s and 80s, industrial design meant creating mass-manufactured products. The prevailing question was *How do you make the perfect cup or chair or sports car?* The challenge was clear: make better, more efficient, more beautiful stuff.

That kind of design still exists, of course, but now the challenge at hand is much more complicated. Why? Because the things we interact with are leaving the simple and physical world and becoming virtual and complex. If industrial design is the interface between the world of technology (defined broadly, to mean anything we design and make) and the world of people, then it has to solve for a whole new level of abstraction. Today, industrial design is as much about designing systems and software and applications as it is about designing objects. We are designing machines, but also the ghosts that live inside them.

A recent IDEO project reflects how much the practice of industrial design has had to stretch in the last couple of decades. For Innova [Schools] (http://www.innovaschools.edu.pe/), the authors were tasked with reconceiving an entire school system in Peru. The brief included many classic design problems—plans for a campus and classroom—but the authors were also faced with creating a business model, a curriculum, and digital tool for educators. Moreover, everything the authors created had to be confluent within a culturally specific system with its own set of requirements. The work still fits my definition of industrial design as a mediation between technology and people, but because it applies across an entire network of information and activities, it will have far more impact than simply designing a better school desk.

That kind of project is what excites me about being a designer today, and it's why I was surprised that the categories hiring the most industrial designers in the NEA [industrial design] report (automotive, for example) were so traditional. That there wasn't a single service company in most of those tables seemed very 20th-century to me. I'd argue that the opportunity for design—where it's taking the greatest strides—is likely to be outside the categories we'd identify in the past. Perhaps the data simply hasn't caught up with the reality of the situation.

In: U.S. Industrial Design Sector
Editor: Louis H. Nielsen

ISBN: 978-1-62948-371-9

Chapter 1

VALUING THE ART OF INDUSTRIAL DESIGN: A PROFILE OF THE SECTOR AND ITS IMPORTANCE TO MANUFACTURING, TECHNOLOGY, AND INNOVATION*

Bonnie Nichols

PREFACE

Design is a field with a large and extensive presence in our nation's manufacturing and services industries, as documented by the national datasets that provide the basis for this report. Designers are prolifically inventing new products, processes, and systems that have a profound impact on our economy and civil society.

The National Endowment for the Arts (NEA) Design Program has been tracking numerous trends in the field of design, from the growing movement of design thinking to social impact design. Although this report brings together, for the first time, analytical perspectives regarding federal data on industrial design, it cannot be all-encompassing. This preface has benefited from conversations with some of the nation's leading designers, design curators, and design firms to convey information not captured by the report itself.

* Published August 2013 by the National Endowment for the Arts, Research Report #56.

An Expanding Definition for Industrial Design

Much of the data in this report are drawn from the U.S. Bureau of Labor Statistics (BLS). The *Occupational Handbook Outlook* from the BLS defines industrial designers as "those individuals that develop the concepts for manufactured products such as cars, home appliances, and toys." While this dataset enables a quantitative grasp of the industry, BLS' definition is arguably limited in scope.

Today's industrial designers find themselves in a variety of roles and functions beyond the development of manufactured products.

Industrial designers are working on projects for a variety of organizations, from government entities to private enterprises. Using a creative lens for approaching complex problems or challenges (often referred to as the design process), designers are engaged by a range of clients to bring a fresh approach to age-old issues.

Industrial designers are not just designing commercial products, but designing user experiences, processes, and systems by applying the creative approach of what has been come to be known as "design thinking." The idea to utilize the design process as a way to analyze and innovate has been widely embraced—from business schools to major consulting practices— and has changed the landscape of how industrial designers work.

For example, an industrial designer might not only design a high-tech medical device for a hospital, but also the patient's interactive experience and touch points with medical staff in the emergency room. Similarly, industrial designers might work with retail merchandisers to reorganize store floor plans and re-imagine the in-store experience for potential customers. Design thinking requires industrial designers to work on diverse teams to solve these more complex challenges. In a typical firm, a team might include an engineer, design strategist, marketer, and anthropologist, as well as software designers and developers, as products become more intelligent and responsive to media inputs.

Historically, industrial design has been a field of invention driven by market demand or clients. Today, there are many examples of designers pursuing clientless endeavors; some have even been on display at the Museum of Modern Art in New York. These artists probably would self-identify as industrial designers; their contributions to developing clientless work as an art form is an acknowledged practice, but one not quantifiable from the datasets explored in this report.

Furthermore, industrial designers are applying their skills to projects that achieve a broader social impact, for populations that cannot afford to hire a designer directly.[1] For example, Design that Matters is a 501(c)(3) nonprofit that "creates new products that allow social enterprises in developing countries to offer improved services and scale more quickly."[2] Design that Matters works with industrial designers, manufacturers, and other nongovernmental organizations to develop scalable products that serve low-resource communities around the globe. Winner of the 2012 National Design Award for Corporate & Institutional Achievement, Design that Matters is developing product innovations within the nonprofit sector for significant impact. One such innovation is an affordable phototherapy device to treat jaundiced newborns in South Asian and Sub-Saharan Africa.

Postsecondary educational programs are retooling their curricula to reflect these more fluid approaches to industrial design. Bachelor's and master's programs in industrial design reflect the interdisciplinary nature of industrial design work. Course curriculums at leading design institutions include anthropology, systems design, and entrepreneurship in addition to the traditional manufacturing and design studio course requirements. Universities are also developing hybrid programs that apply design thinking to business and other fields of practice. Professionals trained in these postsecondary programs might find themselves working for a company where their creative skills are engaged and applied to client-driven work or systems improvement. The cross-disciplinary training and expanded professional role of industrial design thus make it difficult to assess the field conclusively by examining federal datasets alone.

Designers as Inventors

This research report sheds light on the significant role that designers play in invention of new products and services. Creativity often begets innovation and invention. According to data from the U.S. Patent and Trademark Office, approximately 40 percent of inventors named on design patents were also named on utility patents. In contrast, among all inventors named on utility patents, only 2 percent were named on design patents.

These data demonstrate that designers are not only crafting new or improved visual ornamental designs for products—they are simultaneously inventing the new and useful products and processes themselves. While this report quantifies the number of design patents obtained annually, it can be challenging to identify the design firm or individual responsible for crafting a new product design. Most often, patents are assigned to the manufacturer or client company that commissions the design work. While industrial designers may be the ones inventing new products, their work is typically client-driven, with the client organization becoming the patent assignee. Nevertheless, there is no doubt that designers play a crucial role in driving the creative process that results in new inventions or ornamental designs.

Designers are also pursuing their creative process and new product designs through entrepreneurial ventures. While most of the data here account for industrial design establishments and industrial designers on payroll, many designers innovate products outside a formal business establishment. In 2011, $9.2 million was pledged on Kickstarter to support design projects, and 319 successful projects were funded.[3] Since their launch in 2009, 4,424 design projects have been launched on Kickstarter, with a request for $87.7 million.[4] Browsing the posted projects in the design category reveals that the majority of proposals are for product design. New funding platforms such as Kickstarter have enabled entrepreneurial designers to obtain capital to explore conceptual ideas and realize new inventions.

Ultimately design is a tool to inspire innovation and influence systems change. Industrial designers are creative professionals who are doing just that. While we are unable to capture all the facets of the industrial design field in this research report, we acknowledge the breadth of the field and the ever-expanding role of industrial designers as creative professionals who spark innovation, commercial products, and positive change. The NEA report begins to quantify the role of industrial designers in the U.S. economy, and should provoke a more sustained and comprehensive inquiry about the importance of industrial design as a field.

Jason Schupbach, Director of Design
Jennifer Hughes, Design Specialist
National Endowment for the Arts

INTRODUCTION

For nearly four decades, the National Endowment for the Arts has used federally collected data to portray the demographic and financial characteristics of artists as workers.

Designers have always been a vital segment of that population, accounting for nearly 40 percent (about 830,000) of all artists, the largest share for any single artist occupation. The category is broad, and it includes fashion, floral, graphic, interior, and set designers as well as merchandise displayers. It also includes commercial and industrial designers, the subject of this report.

The NEA has not previously attempted to provide in-depth information about the subgroups of occupations that make up the design field. But the availability of such data from the U.S. Bureau of Labor Statistics—specifically through a collaboration with state workforce agencies—makes a detailed rendering possible. When it comes to understanding designers in the context of the industries that employ them, moreover, another federal statistical agency can be consulted: the U.S. Census Bureau.

By using a combination of occupational and industry statistics, this report enumerates industrial designers as workers, the sectors that hire them, the earnings of industrial designers and firms, and the states and metropolitan areas where industrial design workers and industries are most prevalent.

Why do we seek a baseline understanding of industrial design as a field? Part of the answer lies in the reward of creating new knowledge. As much research demonstrates, artists as a whole show great fluidity when it comes to standard categories of employment. They typically rely upon multiple jobs, a combination of part-time and full-time work, and they move with ease across the for-profit and nonprofit sectors. Their portfolio careers, while rich and diverse, can prove difficult to track consistently for a clear picture of artists' contributions to the U.S. economy.

With industrial design, as perhaps with design more generally, a further challenge besets cultural researchers. Industrial design is the handmaiden to a wide range of commercial industries. To appreciate the full impact of industrial design workers and their businesses, it is not enough to view this cohort in isolation. One must trace the relationship of design to product manufacturing and to architectural and engineering firms and a host of other service industries.

This exercise is especially worthy at a time when a growing segment of the American public wants to know exactly where the arts can be found in manufacturing, technology, and product innovation.

As for innovation, this report uses a third federal data source—the U.S. Patent and Trademark Office—to represent designers as inventors. The report thus addresses a specific "node" on the system map that accompanies the NEA's five-year research agenda, set forth in the publication *How Art Works* (2011).[5]

That research topic is titled "Societal Capacities to Innovate and to Express Ideas." The present analysis of the industrial design field allows us better to quantify this elusive component of the U.S. arts ecosystem.

Sunil Iyengar
Director of Research & Analysis
National Endowment for the Arts

EXECUTIVE SUMMARY

Section One: Industrial Design as an Occupation

Industrial designers, also known as commercial designers, develop the concepts for a variety of manufactured projects. Most findings in this section come from data collected by the U.S. Bureau of Labor Statistics.

1) There are more than 40,000 industrial designers in the United States.

- Thirty percent, or approximately 12,000 industrial designers, are self-employed. This share is comparable to that for all artists (34 percent), but more than four times greater than the self-employment rate of all U.S. workers (7 percent).
- Among salaried workers, industrial designers number less than graphic designers (191,440), merchandise-display designers (73,490), floral designers (47,110), and interior designers (40,750)—but they earn more than most design workers.[6]

2) Most salaried industrial designers fall into one of two sectors: manufacturing (11,730 workers) or professional, scientific, and technical services (7,570 workers).

- Transportation equipment manufacturing (makers of auto and aerospace parts and vehicles) is one of the best-paying industries for industrial designers. The industry employs about 20 percent of all industrial designers in manufacturing.
- The professional services sector includes a range of industries. Among those retaining industrial designers on staff are specialized design firms (3,990 industrial designers); engineering and architectural firms (2,510); management, scientific, and technical consulting firms (380); and scientific R&D firms (300).

3) Michigan, Rhode Island, Wisconsin, Indiana, and Pennsylvania are the top five states by percentage of industrial designers in the workforce.

- Four out of the top five metropolitan areas, by concentration of industrial designers in the workforce, are in Michigan. Topping the list is the exception—Kokomo, Indiana, also an automotive manufacturing site.
- California and Michigan each employ more than 3,000 industrial designers. Both states host major hubs of auto design (e.g., the GM Technical Center in Warren, Michigan, and the BMW Design Works and the Volvo Monitoring and Concept Center in Ventura County, California).
- Industrial designers in Michigan earn an average salary of $69,870 a year, or $10,000 more than the national average for this occupation. Industrial designers in Silicon Valley are also among the nation's best-paid industrial designers.

4) Over the next several years, a growing number of industrial designers will find work in the professional services sector (e.g., engineering firms and specialized design firms).

- Industrial designer employment in this sector is projected to leap by 29 percent.
- Manufacturers will continue to employ the greatest share of salaried industrial designers. Yet projected declines in manufacturing employment overall will trim the number of industrial designers in this sector.
- Continued growth is projected for overall employment of industrial designers, though the growth rate (+10.5 percent from 2010 to 2020) is somewhat lower than for all occupations as a whole (+14.3 percent).

Section Two: Industrial Design as a Business

As an industry, industrial design refers to business establishments engaged primarily in creating and developing designs and specifications that optimize the use, value, and appearance of manufactured products. Most findings in this section come from data collected by the U.S. Census Bureau.

1) There are 1,579 industrial design business establishments in the U.S., with a total annual payroll of approximately $1.4 billion.

- In 2007, the most recent year for which such data are available, industrial design firms earned more than $1.5 billion in total revenue. About 94 percent came from sales of product design, model design and fabrication, and other industrial design services.
- Other revenue came from clothing and graphic design, and drafting and other specialized services, among other sources.
- Roughly 14 percent (218) of all industrial design firms earned about $113 million by exporting their services. The sales accounted for roughly one-fourth of these firms' total revenue ($468 million).

2) Total revenues earned by the industrial design industry are highly concentrated among the largest firms.

- The four largest industrial design firms generate 11 percent of the industry's total revenue; the 20 largest, 32 percent; and the 50 largest, 45 percent.[7]
- By contrast, the professional services sector as a whole—of which industrial design is a part—relies far less on larger firms. The 50 largest firms in the sector generate only 18 percent of its total revenue.

3) Rhode Island, Oregon, Michigan, New York, and California rank in the top five states by per-capita concentration of industrial design firms. But a ranking by sheer number of such firms would exclude Rhode Island and Oregon and include Florida and Illinois.

- Rhode Island's industrial design firms cluster around Providence's Rhode Island School of Design (RISD).
- Ranked by payroll size of industrial design firms, the top five states are California, New York, Ohio, Illinois, and Texas. In California, annual payroll of industrial design firms (just under $240 million) is more than twice that of New York's industrial design industry.
- The top metro area ranked by concentration of industrial design firms per capita is Greensboro-High Point, North Carolina—often called the "Furniture Capital of the World." North Carolina is home to many

well-known manufacturers of furniture such as Thomasville, Bernhardt, and Broyhill, which are also the top three "assignees" of North Carolina's design patents.

Section Three: Industrial Design as a Product Innovator

Design patents protect the visual characteristics embodied in or applied to an article. A utility patent protects the way an article is used and works, while a design patent protects the way an article looks. Most findings in this section come from data collected by the U.S. Patent and Trademark Office (USPTO).

1) The number of U.S.-awarded design patents per 100,000 population is at an all-time high: seven in 2012, compared with one at the turn of the 20th century.

- Historical data show two growth spurts for U.S.-awarded design patents: 1910 through the mid-1940s, and the late 1980s through the present date.
- Amid the last economic recession, growth in design patents slowed in 2008 and fell by nearly 10 percent the following year.

2) Between 1998 and 2012, the U.S. Patent and Trademark Office awarded 165,108 design patents (12,445 in 2012 alone) to U.S. companies and individuals.

- More than half of design patents (54 percent) awarded in 1998–2012 fall in eight classes:
  - furnishings
  - recordings, communication, or informational retrieval equipment
  - tools and hardware
  - packages and containers for goods
  - equipment for preparing or serving food or drink
  - transportation
  - environmental heating and cooling or fluid handling
  - games, toys, or sports goods

- One out of five of all U.S. design patents protect the unique external appearance of furnishings or systems used for recording, communication, or information retrieval (e.g., smart phones, computer icons, and computer keyboards).

3) Forty-five percent of design patents are awarded to U.S. companies, and 32 percent to foreign firms. The remainder are awarded mostly to U.S. individuals.

- The top ten U.S. companies by number of industrial design patents awarded are:
  - Microsoft
  - Procter & Gamble
  - Nike
  - Goodyear
  - Black & Decker
  - Wolverine World Wide
  - Kohler Company
  - Apple
  - 3M
  - Ford
- The top ten foreign companies by number of industrial design patents awarded are:
  - Samsung
  - Sony
  - Foxconn
  - LG Electronics
  - Panasonic
  - Honda
  - Nokia
  - Toyota
  - Toshiba
  - Canon
- The top ten U.S. states by industrial design patents awarded per capita are:
  - Washington
  - Wisconsin
  - Oregon
  - Rhode Island

- California
- Minnesota-Ohio
- Illinois
- Massachusetts
- New York
- Michigan

4) Industrial designers are also inventors.

- In an analysis of U.S. patents awarded between 1975 and 2010, Alan Marco, the USPTO's acting chief economist, has found that 40 percent of people named on design patents ("designers") over that period (55,000 out of 136,000) were also named on utility patents. By contrast, among the 2.5 million people named on utility patents ("inventors") over the same period, only 2 percent were named on design patents.
- Over the period studied, the mean number of utility patents naming a designer-inventor (a person named on both a design and utility patent) was 7.3. By contrast, the mean number of utility patents naming an inventor who had not been named on a design patent was 3.3.

PART 1. INDUSTRIAL DESIGN AS AN OCCUPATION

In its 2012–2013 edition of the *Occupational Outlook Handbook*, the U.S. Bureau of Labor Statistics (BLS) reports that industrial designers, also known as commercial designers, develop the concepts for manufactured products such as cars, home appliances, and toys. To varying degrees, they combine art, business, and engineering to make products that people use every day.

The following section draws from BLS data to profile trends in industrial design employment, identify the sectors that employ the greatest numbers of industrial designers, and rank earnings and concentration levels of such workers by state and metropolitan area.

In 2010, there were 40,800 industrial designers working in the U.S. Roughly 30 percent were self-employed—a figure comparable to the self-employment rate for artists of all types (34 percent). By contrast, only 7 percent of all U.S. workers were self-employed.

Most of the findings presented in this section, however, concern the 70 percent of industrial designers who are "wage and salaried non-farm workers."

The data source is BLS' Occupational Employment Statistics (OES) program, which provides far more detailed occupational and industry data than other sources, but which necessarily excludes self-employed workers.[8]

Occupational Employment Statistics

Occupational Employment Statistics is a collaborative program between the BLS and state workforce agencies. The OES surveys 200,000 non-farm establishments every six months, taking three years to fully collect the sample of 1.2 million establishments. OES data are used to report employment and wage estimates for about 800 occupations at national, state, metropolitan, and nonmetropolitan levels.

The OES also reports employment and wage estimates by industry. At the national level, estimates are reported for 450 industries, which are categorized by the North American Industry Classification System (NAICS).

Employment and Earnings by Design Occupations, 2012

	Number employed	Annual median wages
Designers:		
Industrial	29,030	$59,610
Fashion	16,560	$62,860
Floral	47,110	$23,810
Graphic	191,440	$44,150
Interior	40,750	$47,600
Merchandise display	73,490	$26,410
Set and exhibit	8,680	$50,300
All others[1]	7,560	$45,330

Note: Figures exclude self-employed designers

[1] "All other designers" include jewelry designers, memorial designers, and designers of ornamental metalwork. Source: Occupational Employment Statistics, Bureau of Labor Statistics, U.S. Department of Labor

Among all design occupations, industrial designers are comparatively few in number. In 2012, the most recent year for which OES data are available, there were 29,030 industrial designers (excluding self-employed workers). They thus make up a smaller number than do graphic designers (191,440), merchandise display designers (73,490), and floral designers (47,110). Still, industrial designers earn more than most workers in other design occupations (annual median wage: $59,610).[9]

It is worth noting that architecture and landscape architecture—two occupations whose skills and services may intersect with those of designers—are excluded from this comparison, as each is defined separately from design in BLS' occupational taxonomy.[10]

Employment by Industry

Industrial designers are concentrated in two sectors: manufacturing and professional services. In 2012, roughly 11,700 industrial designers worked in manufacturing and 7,600 in professional, scientific, and technical services, which includes architectural and engineering services, specialized design services (including industrial design services), scientific research services, and advertising services, among other industries.[11] Combined, manufacturing and professional services employ 65 percent of salaried workers in this occupation.

On the whole, industrial designers working in manufacturing earn less than designers working in professional services—$56,880 versus $63,580 in 2012. This small difference fades, however, when transportation manufacturing is examined, particularly motor vehicle manufacturing.

In 2012, about 310 industrial designers worked in motor vehicle manufacturing. The BLS withheld earnings data for these workers out of concern that information for such a small group might lead to public disclosure of businesses participating in the OES surveys. Therefore, to identify wages earned by industrial designers working in motor vehicle manufacturing, we turn instead to prior-year OES data.

In 2011, those jobs were far more numerous. That year, there were 800 industrial designers employed in motor vehicle manufacturing; their median earnings were $89,410—$28,000 more than median earnings for all industrial designers. (As will be shown later in this section, the nearly 500-employee drop between 2011 and 2012 reflects a broader decline in employment across the auto manufacturing industry.)

Other transportation-equipment manufacturing industries also pay well. According to the 2012 OES data, industrial designers working in aerospace product manufacturing earned an average of $72,000.[12] Designers employed in the manufacture of motor vehicle parts, which includes car seats, interior trim, and steering wheels, earned median wages of $62,930.

Industrial Designer Employment by Sector, 2012

Sector	Number employed	Annual median wages
Manufacturing	11,730	$56,880
Professional, scientific and technical services	7,570	$63,580
Wholesale trade	3,540	$51,770
Retail trade	510	$50,720
Administrative and support and waste management services	400	$60,880
Construction	350	$56,480
Information	230	$56,980
Federal, state, and local government	210	$52,510
Educational services	60	*
Transportation and warehousing	50	$53,450
Arts, entertainment, and recreation	30	$63,450
Management of companies	n/a	n/a
Other services	*	$60,350

* Not reported by the Bureau of Labor Statistics

n/a: Not reported in this table due to likely data anomaly Detail by sector does not add to total employment of industrial designers.

Source: Occupational Employment Statistics, Bureau of Labor Statistics, U.S. Department of Labor.

Industrial Designer Employment by Selected Manufacturing Industries, 2012

Manufacturing industry	Number employed	Annual median wages
Transportation equipment manufacturing	2,430	$64,470
Motor vehicle manufacturing	310	*
Motor vehicle body and trailer manufacturing	280	$58,300
Motor vehicle parts manufacturing	1,170	$62,930
Aerospace product and parts manufacturing	360	$71,780
Other transportation equipment manufacturing	120	$54,520
Machinery manufacturing	1,820	$53,140
Plastics and rubber products manufacturing	990	$52,650
Electrical equipment appliance and component manufacturing	810	$60,680
Furniture and related product manufacturing	770	$48,150
Computer and electronic product manufacturing	570	$67,800
Jewelry and silverware manufacturing	380	$61,120

* Not reported by the Bureau of Labor Statistics in 2012. In 2011 that figure was $89,410.

Detail by sector does not add to total. 11,730 industrial designers are employed in manufacturing. Source: Occupational Employment Statistics, Bureau of Labor Statistics, U.S. Department of Labor.

In the professional services sector, 85 percent of industrial designers are employed in two industries—"specialized design services" and architectural or engineering services. Of the two industries, specialized design (which includes establishments that provide primarily *industrial* design services) employs a greater number of industrial designers—nearly 4,000 in 2012.

Architectural and engineering services, alternatively, employ fewer industrial designers—about 2,500 in 2012. Nearly all of those designers work in engineering—2,310 in 2012.[13] Industrial designers working in architectural or engineering services earned about $10,000 more than designers in the specialized design services industry.

Industrial Designer Employment by Professional Services Industries, 2012

	Number employed	Annual median wages
Specialized design services	3,990	$58,530
Architectural, engineering, and related services	2,510	$69,250
Computer systems design and related services	*	$63,020
Management, scientific, and technical consulting services	380	$67,530
Scientific research and development services	300	$71,500
Advertising, public relations, and related services	*	$60,280
Other professional, scientific, and technical services	40	$37,660

* Not reported by the Bureau of Labor Statistics

Source: Occupational Employment Statistics, Bureau of Labor Statistics, U.S. Department of Labor

Projected Employment Growth, 2010–2020

The BLS' Office of Employment Projections expects the number of employed industrial designers in the U.S. to reach 45,100 by 2020.[14] This estimate, which includes self-employed *and* wage-and-salary designers, represents a 10.5 percent gain from the number of employed industrial designers in 2010. Although positive, the growth rate projected for industrial designer employment is somewhat below the average growth rate of 14.3 percent forecast for all occupations.[15]

The BLS projections call for wage-and-salary industrial designers to remain at 70 percent of all workers in that occupation, with the remaining 30 percent continuing as self-employed. Among *salaried* industrial designers, employment is projected to rise to 31,400 in 2020—a 9.9 percent gain from 2010 employment levels. (Among non-salaried industrial designers, employment is projected to rise by 12 percent, to 13,700 workers.)

Industrial Designer Employment in Selected Manufacturing and Professional Services Industries, 2010 (Actual) and 2020 (Projected)

	2010		2020		2010 – 2020
	Number employed	Percent of occupation	Number employed	Percent of occupation	Percentage change in employment
All industrial designers	40,800	100.0%	45,100	100.0%	+10.5%
Self-employed	12,200	30.0%	13,700	30.4%	+12.0%
Wage and salary workers	28,600	70.0%	31,400	69.6%	+9.9%
Manufacturing	11,900	29.2%	11,500	25.5%	-3.4%
Selected industries: Plastic products	1,000	2.4%	1,200	2.6%	+21.3%
Machinery	1,900	4.6%	1,800	4.0%	-4.6%
Computer and electronic products	500	1.3%	500	1.0%	-15.0%
Transportation equipment (including motor vehicles and aerospace)	2,300	5.5%	2,300	5.1%	+1.7%
Miscellaneous manufacturing	2,000	5.0%	1,700	3.7%	-19.0%
Professional, scientific, and technical services	7,200	17.6%	9,200	20.5%	+28.5%
Selected industries: Architectural, engineering, and related services	3,200	7.9%	4,200	9.3%	+29.8%
Specialized design services	2,800	7.0%	3,600	8.0%	+27.1%

Note: The employment figures shown were taken from BLS estimates rounded to the nearest one-thousand. Consequently, percentage changes in employment are not necessarily evident in the employment figures.

Source: Office of Employment Projections, Bureau of Labor Statistics, U.S. Department of Labor

By sector, the BLS projects more hiring of industrial designers in the "professional services" sector and reduced employment in manufacturing, a pattern that coincides with the trend of long-term declines in overall manufacturing employment.[16] Despite this decline, manufacturing is projected to remain the leading employer of industrial designers.

In 2010, for example, 7,200 salaried industrial designers were employed by professional services. By 2020, the BLS projects that number to rise by 28.5 percent to reach 9,200 designers. Most of this growth is expected to stem from increased hiring in the two professional services industries employing the greatest numbers of industrial designers—architectural and engineering services and specialized design services. Between 2010 and 2020, both industries are projected to add between 800 and 1,000 industrial designers to their payrolls.

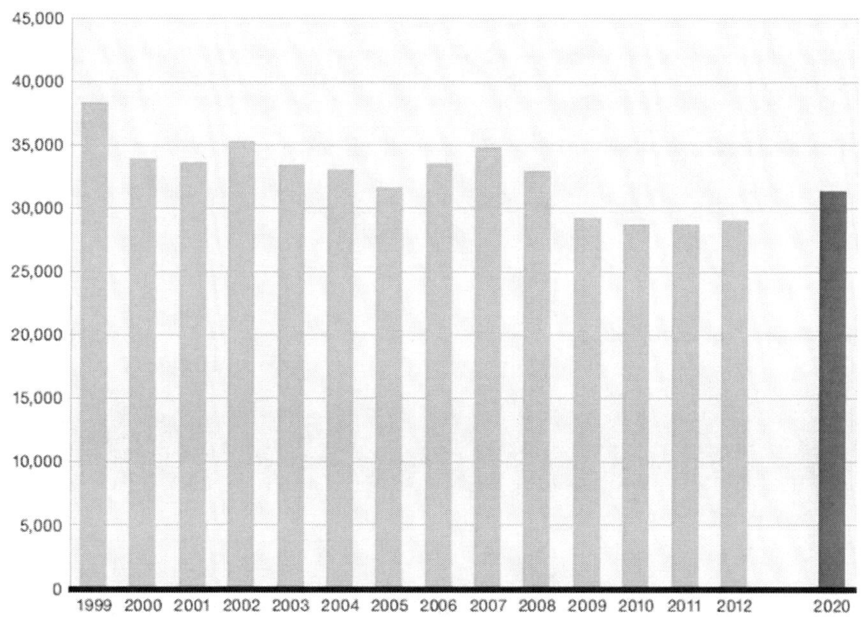

Source: Occupational Employment Statistics and Employment Projections Program, U.S. Bureau of Labor Statistics.

Wage-and-Salary Employment of Industrial Designers, 1999–2012 (Actual) and 2020 (Projected).

Industrial designer employment in manufacturing, as stated earlier, is projected to decline. Between 2010 and 2020, the BLS calls for industrial designer employment in the manufacturing sector to fall from 11,900 workers to 11,500— a dip of 3.4 percent.

Not all manufacturing industries are expected to cut industrial design jobs—the manufacture of plastics, for example, is projected to add a modest number of industrial design jobs, and transportation manufacturing is expected to employ about the same number of industrial designers in 2020 as it did in 2010. However, the BLS projects industrial designer employment to decline in the manufacture of machinery and equipment, computer and electronic products, and "miscellaneous manufacturing," which includes a varied list of manufactured goods such as jewelry, sporting goods and toys, office supplies, and musical instruments.

As noted earlier, manufacturing is projected to remain the largest employer of industrial designers, despite a gradual reduction in the number of workers employed in this sector. In 2020, manufacturing will continue to employ almost 37 percent of salaried industrial designers.

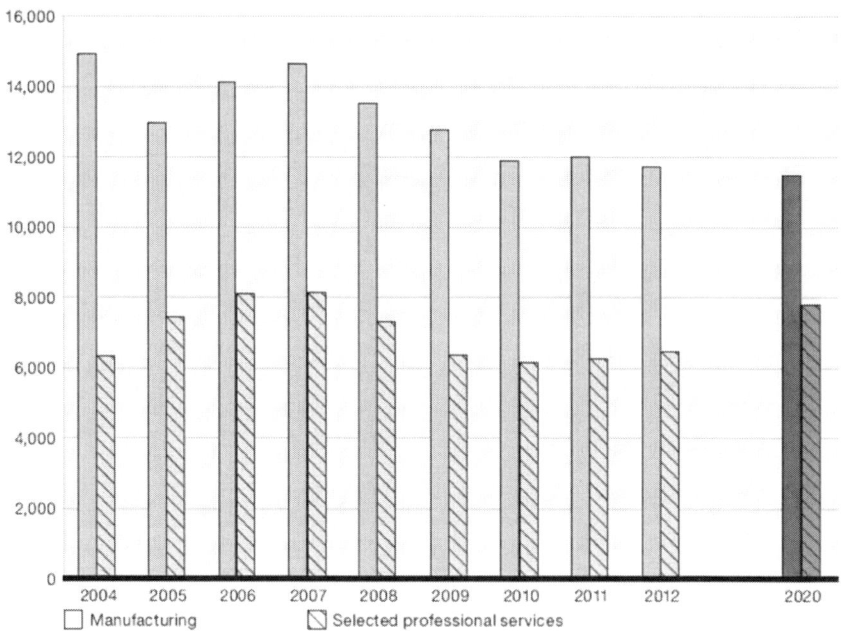

Selected professional services consist of architectural and engineering services and specialized design services.

Source: Occupational Employment Statistics and Employment Projections Program, U.S. Bureau of Labor Statistics

Wage-and-Salary Employment of Industrial Designers, by Manufacturing and Selected Professional Services Industries, 2004-2012 (Actual) and 2020 (Projected).

"One Word: Plastics"

Lower industrial-designer employment in manufacturing coincides with the trend of long-term declines in overall manufacturing employment. An examination of the 25-year period from 1988 to 2012 reveals that employment in the manufacturing sector has steadily declined. In 1988, for example, manufacturing reported 17,906 jobs. By 2012, that number fell to 11,919.

Still, specific manufacturing industries are projected to increase hiring. For example, between 2010 and 2020, the BLS expects plastic manufacturers to add 100,000 jobs. This increased hiring reflects the industry trying to stabilize following deep hiring cuts resulting from the 2007-2009 U.S. recession. In 2000, for example, plastic manufacturers reported nearly 737,000 jobs. In 2010, that count dropped to fewer than 500,000 jobs. By 2020, the BLS expects employment in plastics manufacturing to pick up to almost 600,000 positions.

Construction industries were also hit hard by the recession. Between 2000 and 2010, construction shed almost 1.3 million jobs. But between 2010 and 2020, the BLS projects that construction will add 1.8 million jobs. As construction recovers, it is adding to the employment rosters of the architectural and engineering services industry and of the specialized design services industry, both of which are projected to add industrial designer positions between 2010 and 2020.

Education for Industrial Design

According to the U.S. Bureau of Labor Statistics, industrial designers naturally must have knowledge of design techniques. But they might also need an understanding of engineering and technology, production processes, computers, mathematics, and even physics. Because industrial designers often present their work to marketing staff and clients, the occupation requires English language skills.

A bachelor's degree is required for most entry-level industrial design jobs. As for degrees specific to industrial design, the U.S. Department of Education's National Center for Education Statistics (NCES) reports that U.S. postsecondary schools awarded 1,397 bachelor's degrees in industrial design in 2011, the most recent year for which data are reported.[17] As a share of all bachelor's degrees awarded, industrial design has been just below 1 percent since 2003.[18]

Locations of Industrial Designers

There are only two U.S. states employing more than 3,000 industrial designers— California and Michigan. Michigan, however, far exceeds California in *concentration* of industrial designers. In 2012, as a share of Michigan's employed labor force, industrial designers were nearly four times greater than the U.S. average.[19] The "location quotient" for California's industrial designers, alternatively, was 1.06—as a share of all employed workers in California, industrial designers were 6 percent greater than the U.S. average. Ranked closely behind Michigan is Rhode Island. In 2012, about 320 industrial designers were employed in Rhode Island. Although Rhode Island is home to far fewer designers than California and Michigan, industrial designers compose a large share of Rhode Island's workforce. Relative to all workers in the state, Rhode Island's industrial designers exceed the U.S. average by a factor of three. (*See sidebar on page 31 for more about Rhode Island's industrial design industry.*)

As a share of all employed workers, industrial designers in Wisconsin are twice the national average. Workers in this occupation are also concentrated in Indiana, Pennsylvania, and Ohio, where the location quotient for each state is roughly 1.5, or 50 percent greater than the national average.[20]

States with the Highest Concentrations of Industrial Designers, 2012

	Employed industrial designers	Location quotient
Michigan	3,280	3.75
Rhode Island	320	3.19
Wisconsin	1,800	1.98
Indiana	990	1.57
Pennsylvania	1,880	1.51
Ohio	1,640	1.46
Illinois	1,730	1.38
Iowa	360	1.09
Massachusetts	780	1.09
California	3,370	1.06

Source: Occupational Employment Statistics, Bureau of Labor Statistics, U.S. Department of Labor.

Earnings by State

Not only is Michigan home to large numbers of industrial designers, it also leads in best-paid industrial designers.[21] In 2012, Michigan's industrial designers earned about $10,000 more than the average earned by all U.S. industrial designers. Moreover, Michigan's designers earned a premium of $35,500 above the wages paid to all workers in the state, whose annual earnings averaged $34,750, comparable to the average earned by all U.S. workers.

Contributing to this finding is Michigan's motor vehicle manufacturing industry. As discussed earlier, motor vehicle manufacturing, and related industries such as the manufacture of motor vehicle parts, are among the best-paying industries employing industrial designers. (The U.S. Census Bureau's County Business Pattern data show that the largest annual payroll in motor vehicle manufacturing was in Michigan—$2 billion in 2011.)

Highly paid industrial designers are also found in Oregon and New York, whose 2012 earnings were $69,080 and $67,650, respectively. As discussed in later sections of this report, Oregon and New York, like Michigan, are home to large numbers of industrial design firms.

States with the Best-Paid Industrial Designers, 2012

	Number employed	Median annual earnings	Median annual earnings of all workers	Premium earned by industrial designers
U.S.	29,303	$59,610	$34,750	$24,860
Michigan	3,280	$69,870	$34,370	$35,500
Oregon	200	$69,080	$35,650	$33,430
South Carolina	390	$68,420	$30,210	$38,210
New York	1,810	$67,650	$39,910	$27,740
District of Columbia	30	$67,480	$61,960	$5,520
Massachusetts	780	$65,710	$43,420	$22,290
Connecticut	250	$64,400	$42,030	$22,370
New Jersey	860	$64,160	$39,870	$24,290
New Hampshire	80	$63,800	$35,740	$28,060
Ohio	1,640	$63,210	$33,350	$29,860

Source: Occupational Employment Statistics, Bureau of Labor Statistics, U.S. Department of Labor.

Industrial Designers by Metropolitan Statistical Area

The OES program also provides occupational employment estimates by metropolitan statistical area. In number and concentration of industrial designers, metropolitan areas tend to reflect state rankings. For example, Warren-TroyFarmington Hills, Michigan, employs the greatest number of industrial designers—1,430 in 2012. With 880 employed designers in 2012, Detroit-LivoniaDearborn, Michigan, ranks fourth.[22]

Moreover, five of the top ten metropolitan areas, ranked by concentration of industrial designers, are in Michigan. In addition to the Warren and Detroit areas, this list includes Niles-Benton Harbor, Holland-Grand Haven, and Kalamazoo-Portage.

Metropolitan areas in Wisconsin and Indiana are also on this list, including Wausau and Oshkosh-Neenah in Wisconsin, and Kokomo, Indiana, which ranks first in concentration of industrial designers by metropolitan area. Like the Michigan metro areas on the list, Kokomo is a locus for automotive manufacturing.

Metropolitan Areas with the Highest Concentration of Industrial Designers, 2012

Metropolitan area	Number employed	Location quotient
Kokomo, IN	80	9.19
Niles-Benton Harbor, MI	110	8.36
Holland-Grand Haven, MI	150	6.39
Warren-Troy-Farmington Hills, MI Metropolitan Division	1,430	5.98
Detroit-Livonia-Dearborn, MI Metropolitan Division	880	5.66
Wausau, WI	80	5.59
Anderson, SC	60	4.36
Oshkosh-Neenah, WI	80	4.17
Kalamazoo-Portage, MI	90	3.29
Greenville-Mauldin-Easley, SC	210	3.20

Source: Occupational Employment Statistics, Bureau of Labor Statistics, U.S. Department of Labor

Earnings by Metropolitan Area

Not only does Warren-Troy-Farmington rank highly in number and concentration of industrial designers, but designers working there are also among the best-paid. In 2012, industrial designers working in the Greater Warren-Troy-Farmington area earned annual median wages of $74,730—about $15,000 above average.

Top-Paying Metropolitan Areas for Industrial Designers, 2012

	Num Number employed ber employed	Median annu Median annual earnings al earnings	Median annual earnings of all workers Median annual earnings of all workers	Premium earned by indu Premium earned by industrial designers strial designers
U.S.	29,303	$59,610	$34,750	$24,860
Oxnard-Thousand Oaks-Ventura CA	110	$100,610	$36,930	$63,680
San Jose-Sunnyvale-Santa Clara CA	200	$81,400	$53,470	$27,930
Warren-Troy-Farmington Hills MI Metropolitan Division	1,430	$74,730	$36,150	$38,580
San Francisco-San Mateo-Redwood City CA Metropolitan Division	350	$73,790	$50,690	$23,100
Cincinnati-Middletown OH-KY-IN	570	$73,020	$34,770	$38,250
Portland-Vancouver-Hillsboro OR-WA	200	$71,080	$38,430	$32,650
Dayton OH	100	$70,490	$33,940	$36,550
Santa Ana-Anaheim-Irvine CA Metropolitan Division	350	$69,450	$38,490	$30,960
Greenville-Mauldin-Easley SC	210	$69,440	$30,770	$38,670
Grand Rapids-Wyoming MI	230	$69,350	$33,450	$35,900

Excludes metropolitan areas with fewer than 100 employed industrial designers

Source: Occupational Employment Statistics, Bureau of Labor Statistics, U.S. Department of Labor.

Automotive manufacturing is again shaping the list of areas with well-paid industrial designers. Warren, Michigan, for example, is home to the GM Technical Center, which employs many of GM's designers and engineers. Michigan, however, is not the only automotive design area in the United States. Southern California is home to automotive design centers for most of the major carmakers.

BMW Design Works and the Volvo Monitoring and Concept Center, for example, are located in Ventura County, which is part of California's Oxnard-Thousand Oaks-Ventura metropolitan area. Industrial designers working in this area are the best paid in the country, earning a median annual salary in 2012 of $100,610— $41,000 more than the national average salary for industrial designers. Further, industrial designers working in Oxnard-Thousand Oaks-Ventura earn nearly $64,000 more than the average earned by all workers in the area.

Also well paid are industrial designers working in Santa Ana-Anaheim-Irvine, which counts Ford Advanced Product Creation, Toyota's Calty Design Research, and the Hyundai American Technical Center among its automotive design centers. In 2012, industrial designers working in the Greater Santa Ana-Anaheim-Irvine area earned a median annual salary of $69,450.

Additionally, Silicon Valley's industrial designers earn above-average salaries. In 2012, annual median earnings were $81,400 in the San Jose-Sunnyvale-Santa Clara metro, and $73,790 in Greater San Francisco. Although high by national standards, industrial designer compensation in these areas reflect overall high wages paid in Silicon Valley. In 2012, for example, workers in San Jose-SunnyvaleSanta Clara earned $18,000 more than the average U.S. worker.

Car Design in Southern California—"Detroit West"

Detroit may be the first area that comes to mind as the epicenter of auto design in the U.S. After all, GM, Toyota, and Ford all design cars in the Greater Detroit Metropolitan Area. However, all three companies also have design operations in Southern California, as do many of the world's leading car manufacturers, so much so that the Southern California counties of Ventura, Los Angeles, Orange, and San Diego are collectively called "Detroit West." In addition to design centers for GM, Toyota, and Ford, Southern California is also home to the Volkswagen Design Center (Santa Monica), BMW Design Works (Newbury Park), and Mercedes Advanced Design (Carlsbad), to name just a few.

Southern California's car culture and pleasant temperatures are no doubt a draw to auto designers. But so, too, is the local design talent cultivated by Pasadena's Art Center College of Design. Art Center's alumni include a Who's Who of car design, including Pete Brock and Larry Shindoa, who designed the 1957 Corvette Stingray concept car; Christopher Bangle, BMW's first American chief of design; and Harald Belker, who may be best known for designing vehicles for Hollywood films (e.g., Steven Spielberg's *Minority Report*), but who also helped design the Mercedes Smart Car.

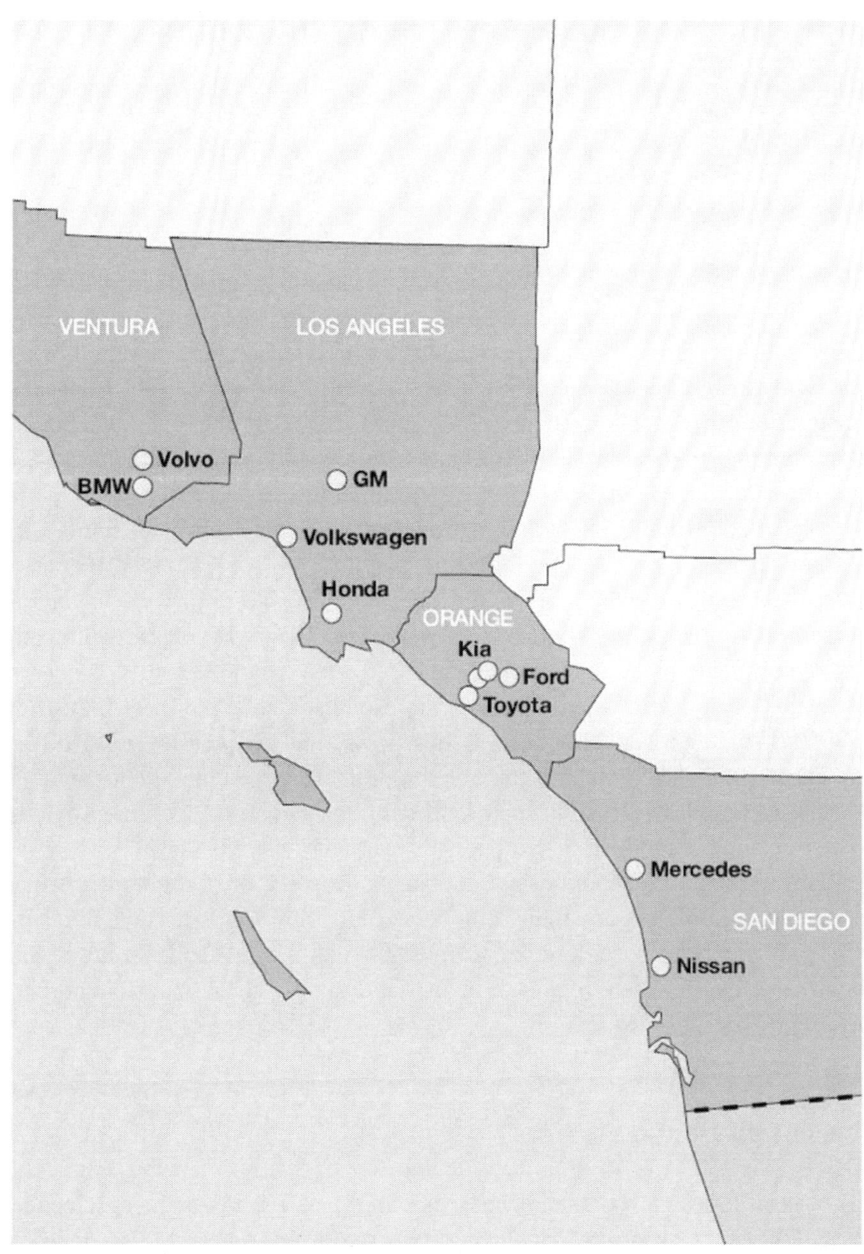

Car Design Centers in Southern California.

PART 2. INDUSTRIAL DESIGN AS A BUSINESS

As discussed in the preceding section, hiring of industrial designers is projected to increase in the professional services sector, which contains a variety of industries, including specialized design services and, more specifically, *industrial design services*.

The U.S. Census Bureau defines industrial or commercial design services, represented by NAICS 541420, as establishments engaged primarily in creating and developing designs and specifications that optimize the use, value, and appearance of their products.[23] The Census Bureau reports that industrial design services can include the determination of the materials, construction, mechanisms, shape, color, and surface finishes of the product, taking into consideration human characteristics and needs, safety, market appeal, and efficiency in production, distribution, use, and maintenance.

This section draws on two U.S. Census Bureau business data sources to examine the industrial design industry. The first is 2011 County Business Patterns (CBP) data, which are based on annual business surveys, including the Service Annual Survey (which surveys 72,000 service businesses with paid employees), and on administrative data extracted from Bureau's Business Register. CBP data are used to enumerate establishments that offer industrial design services, workers employed in these firms, and the industry's payroll.[24]

The second data source for this section of the report is the 2007 Economic Census. Although less current than CBP data, the Economic Census provides greater detail about industrial design firms' services and revenues.[25]

This part of the report relies almost exclusively on Economic Census terminology for sectors and their component industries. Because some industry names are likely to be unfamiliar to most people, a brief explanation of the classification system as it relates to industrial design is given in an appendix to this report (see page 48).

Specialized Design Services

Industrial design establishments make up a small share of the specialized design services industries. However, its payroll is comparable to that of interior design services, an industry employing twice the number of workers.

CBP data show 1,579 business establishments specializing in industrial design in 2011. By contrast, there were 11,135 firms specializing in interior design and 15,000 in graphic design.[26]

The same year, industrial design firms employed 16,308 workers. Although this count was half the number of workers employed by interior design firms, the commercial design services industry's payroll was $1.4 billion, close to the $1.5 billion paid in wages and salaries by interior design firms.

Specialized Design Services, 2011

Specialized design services	Number of establishments	Number of paid employees[2]	Annual payroll ($1,000)
Industrial design services	1,579	16,308	$1,379,490
Graphic design services	15,226	50,241	$2,653,155
Interior design services	11,135	32,478	$1,547,402
Other specialized design services[1]	2,010	8,248	$448,931

[1] Examples include fashion, floats, jewelry, lighting, and textile design services

[2] As of March 12, 20112

Source: County Business Patterns, U.S. Census Bureau, U.S. Department of Commerce.

CBP data also show that the number of industrial design firms and the number of workers employed by them have fluctuated in response to business cycles, particularly during the severe 2007–2009 recession. For example, in 2008, the first full year of the recession, employment in industrial design firms actually grew by more than 28 percent to reach 17,300 workers. But in 2009, the effects of the recession took hold and employment in that industry dropped by 22 percent—the single largest percentage point decline in the 1998–2011 period.[27]

In 2010, the number of workers employed by industrial design firms stabilized at 13,770, then grew sharply in 2011 to reach 16,300 workers, a number suggesting the beginning of a recovery in industrial design employment.[28]

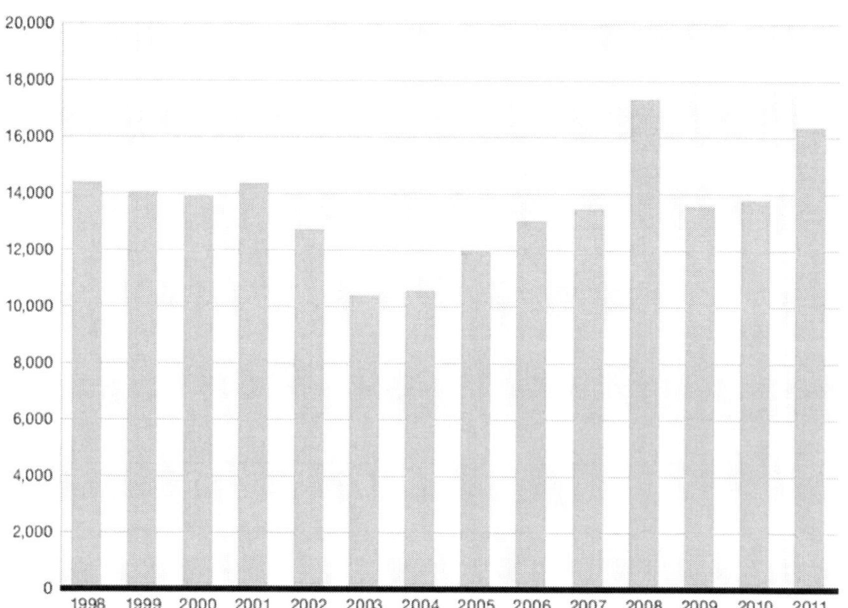

Source: County Business Patterns, U.S. Census Bureau, U.S. Department of Commerce.

Number of Workers Employed in Industrial Design Firms, 1998–2011.

Revenue Sources

Not surprisingly, most of the revenue earned by industrial design establishments stems from sales of product design and related services. The 2007 Economic Census, the most recent business census for which statistics are available, showed that product design and model design and fabrication (as well as "other industrial design") accounted for nearly 94 percent of the revenue earned by industrial design firms.[29] In 2007, product design and related services generated $1.4 billion in sales.

In addition to their primary work of designing products, some industrial design businesses offer other services such as drafting and graphic and clothing design. In 2007, for example, 148 industrial design firms provided graphic design services. Although that number was small—fewer than 10 percent of all the establishments in that industry—graphic design services generated $37.1 million in sales, and half of that figure ($18.8 million) came from sales of graphic designs of corporate or organizational images.

Products and Services Provided by Industrial Design Firms, 2007

	Establishments	Revenue ($1,000)	Percent of industry revenue
Industry total	1,637	$1,541,787	100.0%
Industrial design services	1,605	$1,441,980	93.5%
Product design	1,243	$941,596	61.1%
Model design and fabrication	656	$274,511	17.8%
Other industrial design	457	$217,278	14.1%
Drafting services	25	$3,152	0.2%
Graphic design services	148	$37,142	2.4%
Corporate/organization image graphic design services	79	$18,766	1.2%
Advertising graphic design creative services	57	$9,079	0.6%
Publication graphic design services	19	$375	*
Commercial illustration graphic design services	29	$820	0.1%
Website design and development services	53	$1,270	0.1%
All other graphic design services	17	$6,831	0.4%
Clothing design services	22	$6,515	0.4%
Other specialized design services	11	$10,112	0.7%
Resale of merchandise	60	$22,647	1.5%
All other operating revenue	46	$19,968	1.3%

* Not reported

Source: 2007 Economic Census, U.S. Census Bureau, U.S. Department of Commerce.

Industrial Design Exports

The Economic Census also tracks sales from exported services. In 2007, 218 industrial design firms exported $112.7 million in services, amounting to one-quarter of all revenue earned by firms exporting industrial design services.

Industrial Design Services Establishments Exported Services, 2007

Number of establishments	Total revenue of these	Revenue from exported services
exporting services	establishments ($1,000)	($1,000)
218	$467,900	$112,745

Source: 2007 Economic Census, U.S. Census Bureau, U.S. Department of Commerce.

Revenue Earned by the Largest Industrial Design Firms

While most business data produced by the U.S. Census Bureau are reported for "establishments," the Economic Census also shows the percent of revenue earned by the largest "firms," which refers to a business organization consisting of one domestic establishment or more under common ownership.

That small distinction aside, the Economic Census shows that much of the revenue earned in this industry is concentrated among the top companies, more so than for other firms in the professional services sector, to which industrial design firms belong.[30]

Percent of Revenue Earned by Largest Firms, 2007

	Number of establishments	Revenue ($1,000)	Percent of revenue earned by largest firms
Industrial design services			
All firms	1,637	$1,541,787	100.0%
4 largest firms	11	$176,136	11.4%
8 largest firms	16	$290,131	18.8%
20 largest firms	31	$488,561	31.7%
50 largest firms	73	$693,153	45.0%
Professional, scientific, and technical services			
All firms	847,492	$1,251,003,504	100.0%
4 largest firms	3,067	$52,890,516	4.2%
8 largest firms	7,086	$86,265,201	6.9%
20 largest firms	8,779	$155,028,542	12.4%
50 largest firms	19,898	$229,218,067	18.3%

Source: 2007 Economic Census, U.S. Census Bureau, U.S. Department of Commerce.

In 2007, for example, the four largest industrial design firms (i.e., largest in revenue) earned 11.4 percent of all the revenue earned in that industry. Although the U.S. Census Bureau is barred from reporting data on individual establishments, industry analysts suggest that the top four are Continuum (West Newton, Massachusetts), IDEO (Palo Alto, California), Frog Designs (San Francisco, California), and Lunar Design (Palo Alto, California).

By contrast, the four largest firms in all professional, scientific, and technical services (the sector that encompasses industrial design firms) earned 4.2 percent of the sector's revenue.

Comparing industrial design firms to other specialized design companies underscores this concentration. In 2007, the top 50 industrial design services firms earned 45 percent of the industry's revenue, while the top 50 interior and graphic design companies earned 12–13 percent.

The top 50 "other specialized design" firms earned 39 percent of the industry's revenue—above average but below the share reported for the top industrial design firms.

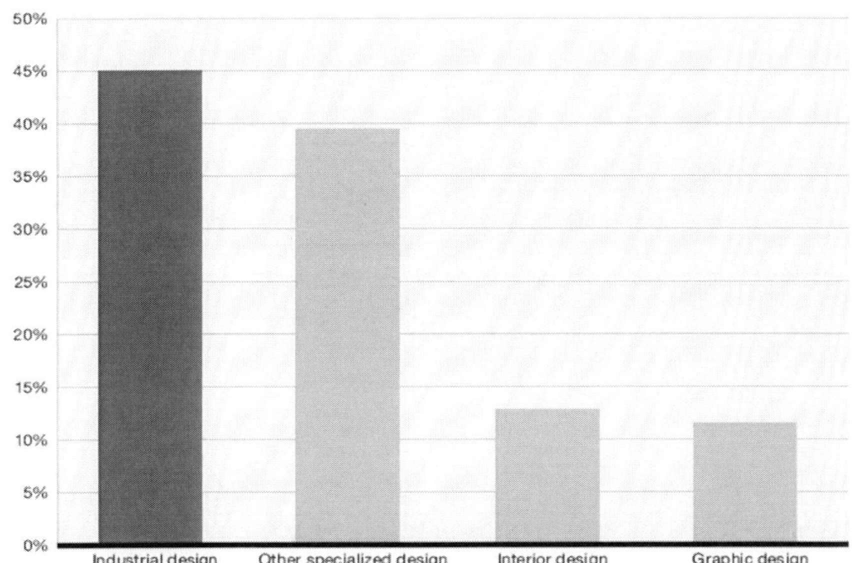

Source: 2007 Economic Census, U.S. Census Bureau, U.S. Department of Commerce.

Percent of Revenue Earned by the 50 Largest Specialized Design Firms, per Industry, 2007.

State Location of Industrial Design Businesses

Returning to CBP data, state tabulations show that the greatest numbers of industrial design services establishments are in the large-population states of California, New York, Florida, Illinois, and Michigan. In 2011, those five states were home to half of all U.S. industrial design firms.

These states also rank highly in numbers of industrial design establishments per capita. In California, New York, and Michigan, for example, there were roughly four establishments per 500,000 population—a figure well above the U.S. average of 2.5. However, in number of industrial design establishments per capita, no state ranks as high as Rhode Island. In 2011, 18 establishments were located in Rhode Island. Given the state's 2011 population of 1.1 million, Rhode Island, consequently, was home to 8.6 industrial design firms per 500,000 population.

Industrial Design Services Establishments, 2011 Selected Top States

	Number of establishments	Per capita1
U.S.	1,579	2.5
California	299	4.0
New York	168	4.3
Florida	127	3.3
Illinois	95	3.7
Michigan	85	4.3
Ohio	74	3.2
North Carolina	72	3.7
Massachusetts	49	3.7
Oregon	34	4.4
Connecticut	26	3.6
Rhode Island	18	8.6

1 Per 500,000 population

Source: County Business Patterns, U.S. Census Bureau, U.S. Department of Commerce.

Design in Rhode Island

Per capita, Rhode Island is home to more industrial design establishments than any other state. In 2011, there were roughly eight industrial design firms per 500,000 population—a count significantly higher than the national average of 2.5, and higher than the four establishments per 500,000 population in California, New York, or Michigan. Rhode Island's primacy in industrial design services stems in part from the Rhode Island School of Design (RISD) in Providence. There are 18 industrial design firms in Rhode Island, ten in Providence alone. Examples of local firms are Observatory, Tellart, Ximedica, Item New Product Development, and Fuzion Design. Notably, toy manufacturer Hasbro is headquartered in nearby Pawtucket, Rhode Island.

RISD was founded by Helen Rodelia Rowe Metcalf in 1877—more than 40 years before women gained the right to vote. Its alumni include Stuart Karten, founder of the Los Angeles-based industrial design firm, Karten Design, and David Hanson, well known for his creation of realistic, humanoid robots.

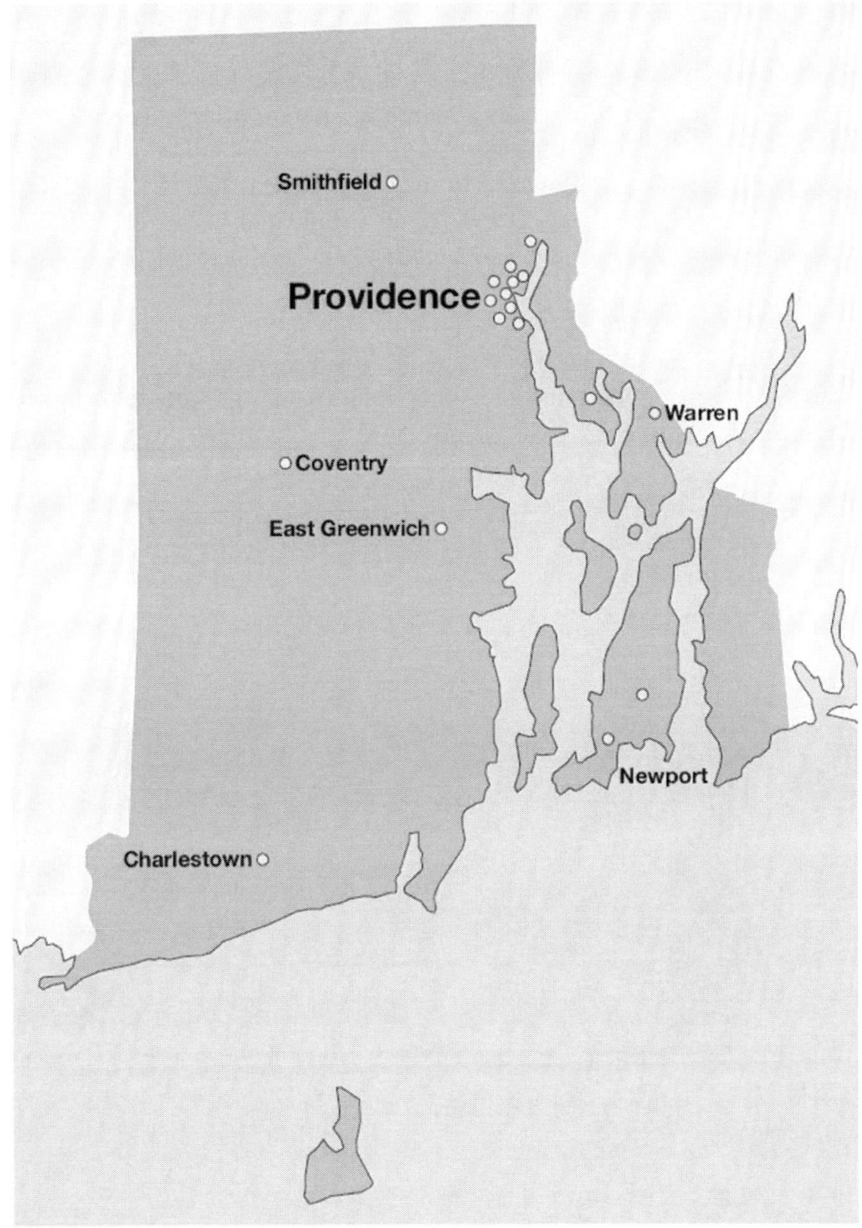

Source: County Business Patterns, U.S. Census Bureau, U.S. Department of Commerce.

Industrial Design Services in Rhode Island Business Establishments in 2011.

State Payroll

While CBP data do not report revenue, they do show payroll, which is used here to gauge the finances of industrial design establishments by state. In 2011, the payrolls of California's industrial design firms exceeded $239 million, and in New York, the industry's payroll was $94 million. Other states ranking high in industrial design firms' payroll included Ohio ($86.7 million), Illinois ($74.1 million), and Texas, which, while home to only 59 such establishments in 2011, generated $42 million in industrial design payroll.

California's large industrial design payroll reflects the sheer size of the industry in that state. For example, California's payroll for the industry is more than two times greater than for New York's industrial design sector. Also, California is home to twice the number of industrial design firms and workers as New York.

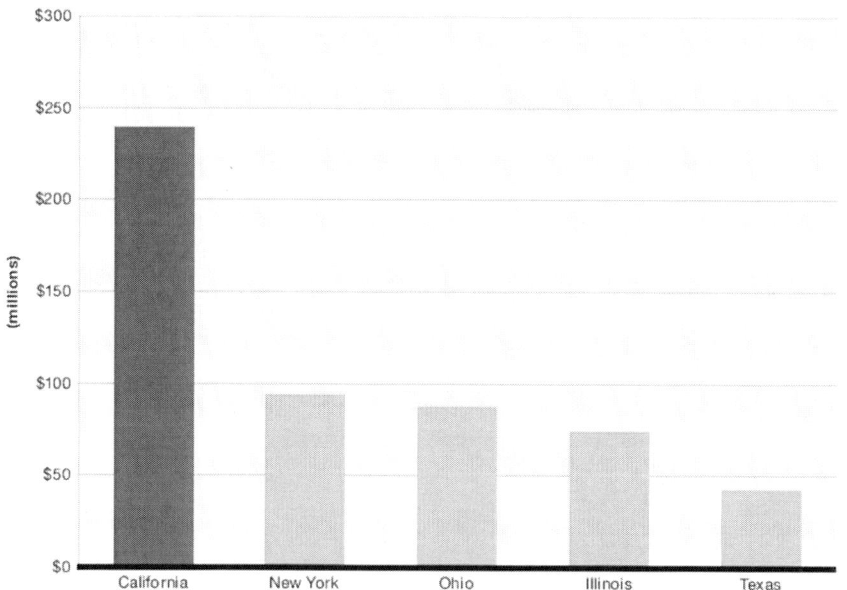

Source: County Business Patterns, U.S. Census Bureau, U.S. Department of Commerce.

Industrial Design Services Payroll Top Five States in 2011.

Industrial Design Businesses by Metropolitan Statistical Area

CBP data also show counts of business establishments in the industrial design services industry for metropolitan statistical areas, though estimates of payroll and number of employees are typically reported for large metro areas only.

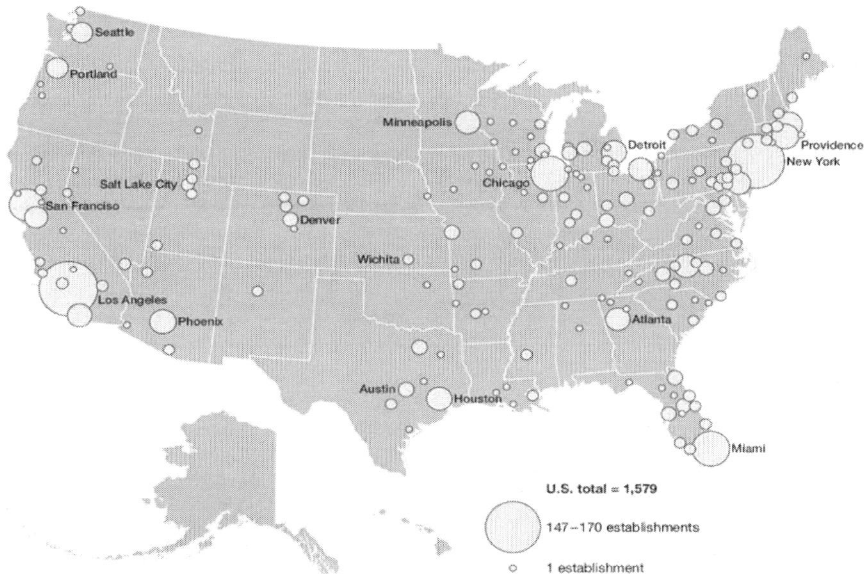

Source: County Business Patterns, U.S. Census Bureau, U.S. Department of Commerce.

Industrial Design Services; Business Establishments in 2011.

In 2011, the New York and Los Angeles metropolitan areas scored the greatest numbers of establishments: 170 and 147, respectively. In fact, 20 percent of all U.S. industrial design establishments are located in either the New York and Los Angeles areas. These two areas also rank highly in number of firms per capita. In New York, there were 4.3 industrial design establishments per 500,000 population, and, in Los Angeles, even more—5.7 per 500,000 population. Per capita, the San Francisco, California area had 6.9 industrial design firms in 2011, while Portland (Oregon), San Jose (California), and Providence (Rhode Island) each had 6.2 per 500,000 population. However, no metro area approaches Greensboro-High Point (North Carolina) in number of establishments per capita. In 2011, the estimated population for Greensboro-High Point was just over 700,000, giving the area 21.2 industrial design firms per 500,000 population.

Industrial Design Services Establishments, 2011 Top Metropolitan Areas

	Number of establishments	Per capita[1]	Payroll ($1,000)	Number of employees
U.S.	1,579	2.5		
New York-Northern New Jersey-Long Island, NY-NJ-PA	170	4.3	$87,053	1,183
Los Angeles-Long Beach-Santa Ana, CA	147	5.7	$92,454	1,030
Chicago-Naperville-Joliet, IL-IN-WI	86	4.5	$70,253	700
San Francisco-Oakland-Fremont, CA	61	6.9	$64,624	580
Miami-Fort Lauderdale-Pompano Beach, FL	56	4.9	$5,838	124
Detroit-Warren-Livonia, MI	39	4.5	D	a
Boston-Cambridge-Quincy, MA-NH	37	4.0	$21,126	260
Greensboro-High Point, NC[2]	31	21.2	$6,245	114
San Diego-Carlsbad-San Marcos, CA	29	4.6	$10,444	157
Portland-Vancouver-Beaverton, OR-WA	28	6.2	$12,517	207
Minneapolis-St. Paul-Bloomington, MN-WI	27	4.0	$9,588	127
Cleveland-Elyria-Mentor, OH	26	6.3	$16,227	287
Seattle-Tacoma-Bellevue, WA	26	3.7	$30,690	360
San Jose-Sunnyvale-Santa Clara, CA	23	6.2	D	b
Providence-New Bedford-Fall River, RI-MA	20	6.2	$10,222	129

[1] Per 500,000 population

[2] Population in 2011 was 730,531

D: Withheld to avoid disclosing data for individual companies

2,500–4,999 employees

250–499 employees

Note: 20 industrial design services establishments were also reported for the Philadelphia, Houston, Atlanta, and Phoenix metropolitan areas. However, each of these metros was at or below the national per capita average of 2.5.

Source: County Business Patterns, U.S. Census Bureau, U.S. Department of Commerce.

Propelling Greensboro-High Point's per-capita count of commercial design firms is the area's furniture industry—indeed, High Point has been called the "Furniture Capital of the World." As will be discussed in Part 3 of this report, furnishings account for a large share of design patents awarded by the U.S. Patent and Trademark Office (USPTO). U.S. design patents awarded in North Carolina are dominated by the furnishings product class and by well-known furniture manufacturers such as Thomasville, Bernhardt, and Broyhill, which are the top three "assignees" of North Carolina's design patents.[31]

PART 3. INDUSTRIAL DESIGN AS A PRODUCT INNOVATOR

So far, this report has profiled the industrial design field in terms of workers and establishments, their locations and earnings, and their actual and projected growth rates. But what does industrial design contribute to product innovation? To answer the question, we turn to an additional metric: the annual growth rate of U.S.-awarded design patents. Patents are not necessarily a perfect measure of product innovation, but they are an accessible and concrete measure.

Through this lens, we witness the 19th-century birth of U.S. industrial design and ensuing periods of heightened design activity such as the 1930s era of streamlining. As of 2012, moreover, there were roughly seven U.S.-awarded design patents per 100,000 population—an all-time high for industrial design in this country.

Patent data also illustrate the breadth of products designed by industrial designers. Patents protect the designs of products ranging from ice cream cones and dog beds, to power tools and high chairs. Design patents were at the center of the recent billion-dollar legal dispute between rival smartphone manufacturers Apple and Samsung.

Patent data show that industrial design is big business. In addition to Apple and Samsung, companies such as Microsoft and Nike lead in U.S. awards of design patents. The headquarters of U.S. manufacturing firms, in turn, contribute to a geographic pattern in which design-patent awards are concentrated in western and midwestern states.

Finally, patent data reveal that industrial designers are inventive. "Designers-inventors" (individuals who are named on both design *and* utility patents) are named on substantially more utility patents than inventors named only on utility patents.

A Brief History of Design Patents

Although the U.S. government awarded its first patent in 1790 (signed by President George Washington and Secretary of State Thomas Jefferson), patents initially protected only the invention of a new article or the improvement of a processing system. Patents to protect the *design* of a "useful" product were not available until the 1840s, when new manufacturing technologies introduced design elements as a means of creating distinctive consumer products. In particular, the manufacture of two products, iron stoves and patterned textiles, illustrate the development of U.S. industrial design and the motive for protecting designs through patents.

In the 1830s, New York's Jordan Mott revolutionized the process for manufacturing cast-iron stoves by incorporating fluting and curves into his stove designs. First added to enhance heat dissipation, these design elements distinguished Mott's stoves and created consumer appeal for the stoves and other metal products he manufactured. During this same period, New England's textile manufacturer Francis Lowell designed an ornate calico print that became so popular it largely replaced checks and home-spun plaids.

Because design patents were unavailable in the 1830s, both Mott and Lowell were subject to piracy from their competitors until 1842. That year, with help from Mott's lobbying efforts, the U.S. passed a statute providing for the grant of patents for "any new and original design for a manufacture or for printing on fabric."

Today, the U.S. Patent and Trademark Office defines "design" as the visual characteristics embodied in or applied to an article. The design patent relates to the configuration or shape of an article, to the surface ornamentation applied to an article, or to the combination of the configuration and surface ornamentation. To distinguish between design and utility, the USPTO guide notes that a "utility" patent protects the way an article is *used* and *works*, while a "design" patent protects the way an article *looks*.

Forms of Intellectual Property Protection

Utility patents protect a new and useful process, machine, article of manufacture, composition of matter, or any new and useful improvement thereof. The statutory period is 20 years after the application date.

Design patents protect new, original, and ornamental design for an article of manufacture. The statutory period is 14 years from date of issuance.

Plant patents protect the invention or discovery of asexually reproduced new plant variety. The statutory period is 20 years after the application date.

Copyright protects the authors of literary, dramatic, musical, artistic, and other intellectual works, both published and unpublished. The copyright protects the form of expression rather than the subject matter of the work. Per the Copyright Act of 1976, copyrighted works created as of January 1978 are protected for the life of the author plus an additional 70 years.

Trademark laws protect a word, name, symbol, device, or design and packaging (as in trade dressing) that is used in trade with goods to indicate the source of the goods and to distinguish them from the goods of others. Trademark rights prevent others from using a confusingly similar mark or design/packaging. Trademark rights apply until abandonment.

Trade secret laws protect technological and industrial information not generally known in the trade against unauthorized industrial use by others. Legal protection is afforded only to owners who have taken diligent steps to preserve trade secrets.

Types of Designs Protected by Patents

Examining design patents by class reveals the diversity of industrial design. U.S. patent designs are awarded in product-design classes ranging from food items (e.g., the design of chocolate bars) and animal husbandry products (such as horse saddles and bird houses), to the design elements of dishwashers, swimming pools, and musical instruments.

Of the 32 distinct design patent classes, however, more than half of all design patents granted in the U.S. are awarded in just eight classes, with the top two classes accounting for 20 percent.[32] Between 1998 and 2012, the USPTO awarded 29,540 design grants in furnishings—a class covering the design of furniture, cabinets, and carpets—and in recording, communication, and information equipment, which includes the design of smartphones, computer icons, and computer keyboards.

The remaining top classes include design patents in tools and hardware, packages and containers, transportation, and toys.

U.S.-Awarded Design Patents, Top Eight Classes, 1998–2012

Class	Class Title	1998–2012	2012	Share of all design patents: 1998–2012
ALL	All design patents	165,108	12,445	100.0%
	Design patents granted in top eight classes	89,542	6,457	54.2%
D06	Furnishings *(Cabinet, rocking chair, playpen, carpet design)*	16,393	1,103	9.9%
D14	Recording, communication, or information retrieval equipment *(Mobile telephone, mouse pad, computer icon, barcode scanner)*	13,147	1,194	8.0%
D08	Tools and hardware *(Chain saw, furniture castor, shelf brackets, scissors)*	11,142	696	6.7%
D09	Packages and containers for goods *(Perfume bottle, egg carton, drinking bottle)*	10,164	867	6.2%
D07	Equipment for preparing or serving food or drink[1] *(Espresso maker, banana holder, napkin holder, pizza oven)*	10,019	858	6.1%
D12	Transportation *(Car chassis, golf cart, wheelchair, ATV)*	9,991	579	6.1%
D23	Environmental heating and cooling; fluid handling *(Outdoor fountain, sink faucet, sauna, ceiling fan)*	9,449	576	5.7%
D21	Games, toys, and sports goods *(Hockey stick, treadmill, baby rattle, jukebox)*	9,237	584	5.6%

[1] Not elsewhere classified

Data source: U.S. Patent and Trademark Office, U.S. Department of Commerce.

Design Patents by Ownership

Companies, both U.S. and foreign, receive roughly 75 percent of the design patents granted by the USPTO. In 2012, that split was 45 percent to U.S. companies and 32 percent to foreign businesses.[33]

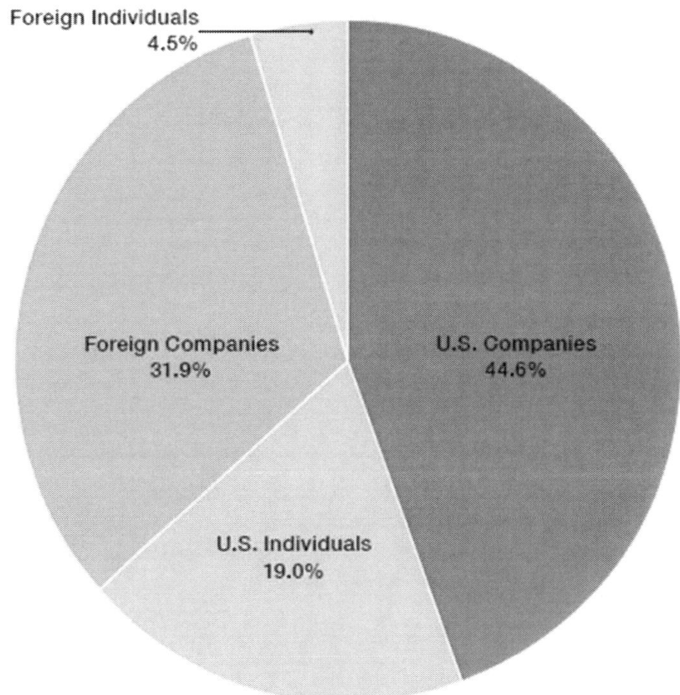

Data source: U.S. Patent and Trademark Office, U.S. Department of Commerce.

U.S.-Awarded Design Patents by Ownership, 2012.

Examining design patents granted to companies by country reveals that designed manufactured goods (manufactured goods protected by U.S. design patents) sold in this country are dominated by a diverse U.S. manufacturing industry and by foreign manufacturers concentrated among Asian electronics and auto industries.

In number of U.S. design patents granted, the top ten U.S. companies represent a wide range of manufacturing. First on this list is Microsoft, followed by Procter & Gamble and Nike. Each of these companies, which collectively represent such diverse sectors as software, consumer products, and sporting apparel, were granted more than 1,500 design patents between 1999–2012.

The top U.S. list also includes manufacturers of tires and rubber (Goodyear), plumbing fixtures (Kohler), office supplies (3M), and auto manufacturing (Ford). Also on the list, Apple, maker of the iPhone and iPad, received 788 U.S. design patents from 1999 to 2012.

U.S.-Awarded Design Patents, 1999–2012

U.S. Headquarter Companies—Top 10 in Design Patents

Grantee	Design patents granted:1999 –2012	Headquarters	Industry
All U.S. headquartered companies	101,376	n/a	n/a
Microsoft Corporation	1,900	Redmond, WA	Computer software/video games
Procter & Gamble Company	1,842	Cincinnati, OH	Personal care
Nike, Inc.	1,690	Washington County, OR	Footwear and apparel
Goodyear Tire and Rubber Company	1,260	Akron, OH	Tires and rubber
Black & Decker, Inc.	1,023	Towson, MD	Power tools/hardware
Wolverine World Wide	939	Rockford, MI	Footwear
Kohler Company	853	Kohler, WI	Plumbing, cabinetry, engines
Apple, Inc.	788	Cupertino, CA	Computer hardware/software
3M Innovative Properties Company	754	St. Paul, MN	Diversified (e.g., Post-It, Thinsulate)
Ford Motor Company[1]	745	Dearborn, MI	Automotive

[1] Includes design patents awarded to Ford Global Technologies

Data source: U.S. Patent and Trademark Office, U.S. Department of Commerce.

Among foreign companies granted U.S. design patents, Asian countries, particularly companies headquartered in Japan and Taiwan, compose the largest share. Moreover, the Asian companies granted U.S. design patents are concentrated in electronics and auto manufacturing. Between 1999 and 2012, Sony was granted more than 3,000 U.S. design patents. Honda received more than 1,200 U.S. design patents.

Although companies in the United States are granted more U.S. design patents than foreign-owned companies, the top ten foreign companies in number of design patents receive more patents than U.S. companies ranking highly in awards of design patents. Between 1999 and 2012, the top ten foreign companies in grants of U.S. design patents received 15,561 patents. Over this same period, the top ten U.S. companies received 11,794 design patents. This concentration—a large number of U.S. design patents granted to fewer foreign companies—stems largely from Samsung and Sony. Between 1999 and 2012, each company received more than 3,000 U.S. design patents.

U.S.-Awarded Design Patents, 1999–2012

Foreign-Headquartered Companies—Top 10 in Design Patents

Grantee	Design patents granted: 1999 –2012	Headquarters	Industry
All foreign-headquartered companies	87,433	n/a	n/a
Samsung Electronics Co.	3,323	South Korea	Electronics and information technology
Sony Corporation	3,308	Japan	Electronics and entertainment
Foxconn (Hon Hai Precision Industry Co.)	1,613	Taiwan	Electronics
LG Electronics	1,496	South Korea	Diversified (e.g., electronics, entertainment)
Panasonic[1]	1,488	Japan	Electronics
Honda Motor Company	1,239	Japan	Automotive
Nokia Corporation	1,051	Finland	Communications and information technology
Toyota Motor Corporation	973	Japan	Automotive
Toshiba Corporation	588	Japan	Engineering and electronics
Canon, Inc.	482	Japan	Imaging and optical products

[1] Includes design patents granted to the Matsushita Industrial Company in 1999 through 2008.

Data source: U.S. Patent and Trademark Office, U.S. Department of Commerce

The Patent Trial of the Century: ***Apple vs. Samsung***

In August 2012, a U.S. district court awarded Apple more than $1 billion in damages in a patent dispute with rival smartphone and tablet manufacturer Samsung. Although the award to Apple was reduced to $599 million in March 2013, it remains among the highest patent-infringement awards. *The Wal Street Journal* termed the dispute "The Patent Trial of the Century."

The case was particularly relevant to industrial design because Apple and Samsung's arguments hinged on design rights—four of the six patents contested were design patents.

Underscoring the importance Apple has placed on design, *InformationWeek* reported in September 2006 that 11.8 percent of Apple's patent portfolio is made up of design patents—a share well above the electronics industry's average of 2.7 percent.

State Clustering of Design Patents

A geographic analysis of design patents reinforces the general impression that much of U.S. industrial design is clustered in western and midwestern states.

For example, California ranks first and foremost in grants of U.S. design patents. Between 2008 and 2012, 12,520 design patents were awarded in California. This tally represents 20 percent of all state-level design patents. Ranking a distant second was New York, which garnered roughly 4,600 design patents, or 7.3 percent of the state total.

Adjusting for state population, however, changes that ranking. Per capita, Washington state ranks first in number of U.S. design patents granted between 2008 through 2012. Over this period, Washington garnered 40.3 design patents per capita. Washington State is also home to Microsoft Corporation, which is among the top-ranking companies in design patent awards.

U.S.-Awarded Design Patents, Selected Top States, 2008–2012

State	Per capita design patents granted[1]	Illustrative Corporate Headquarters
Washington	40.3	Microsoft Corporation
Wisconsin	39.0	S.C. Johnson & Son, Inc.
Oregon	38.8	Nike, Inc.
Rhode Island	34.5	Hasbro, Inc.
California	33.5	Apple, Inc.
Minnesota, Ohio	32.7, 32.5	3M Innovative Properties Company, Procter & Gamble Company
Illinois	29.0	Motorola, Inc.
Massachusetts	25.5	Gillette Company
New York	23.7	Colgate-Palmolive Company
Michigan	22.8	Ford Motor Company

[1] Per 100,000 people, calculated from U.S. Census Bureau population estimates for 2008–2012.

Data source: U.S. Patent and Trademark Office, U.S. Department of Commerce.

Following closely in state rank are Wisconsin and Oregon, each granted 39 design patents per capita between 2008 and 2012. Wisconsin and Oregon are home to S.C. Johnson & Son and Nike, respectively. Over this five-year period, 34.5 design patents, per 100,000 population, were awarded in Rhode Island, home to the Hasbro toy company.

In addition to Wisconsin, other Great Lake states ranking highly in numbers of design patents include Minnesota and Ohio. Between 2008 and 2012, 32.7 design patents, per capita, were awarded in Minnesota. Minnesota is headquarters to the 3M Innovative Properties Company (maker of Post-It and Thinsulate). Over this period, nearly the same number of design patents, per 100,000 people, were awarded in Ohio, home to both Procter & Gamble and the Goodyear Tire and Rubber Company.

Although the western and midwestern states dominate in awards of U.S. design patents, Massachusetts and New York also rank highly with roughly 25 design patents per capita. Contributing to this ranking are New York's Colgate-Palmolive Company and the Gillette Company in Massachusetts.

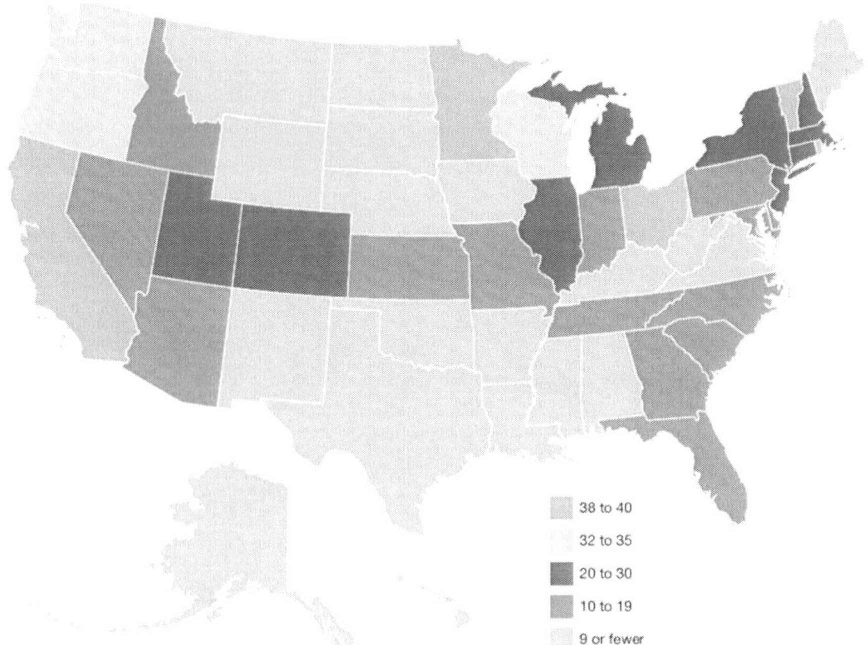

Data Source: U.S. Patent and Trademark Office, U.S. Department of Commerce.

U.S. Design Patents Granted, 2008–2012; Per Capita (100,000 People).

Business Cycles and Trends in Design Patents

The number of U.S. design patents ebbs and flows with U.S. business cycles. In 1999, the USPTO awarded 14,732 design patents (both U.S. and of

foreign origin). Although there was strong growth in the number of design patents granted in 2000, the relatively mild U.S. recession of 2001 likely contributed to the decline in U.S. design patents granted in 2001 and 2002. As the U.S. economy recovered, there was an increase in the number of design patents awarded. For example, USPTO-awarded design patents grew by nearly 15 percent in both 2006 and 2007.

In December of 2007, however, the U.S. economy fell into a severe recession that lasted until June 2009. Growth in design-patent awards slowed to 6 percent in 2008, followed by a sharp drop in design patents of nearly 10 percent in 2009. The number of design patents granted continued to fall, albeit less sharply, in 2010 and 2011. Recent counts for 2012 show a modest recovery of 2.8 percent growth in U.S.-awarded design patents.

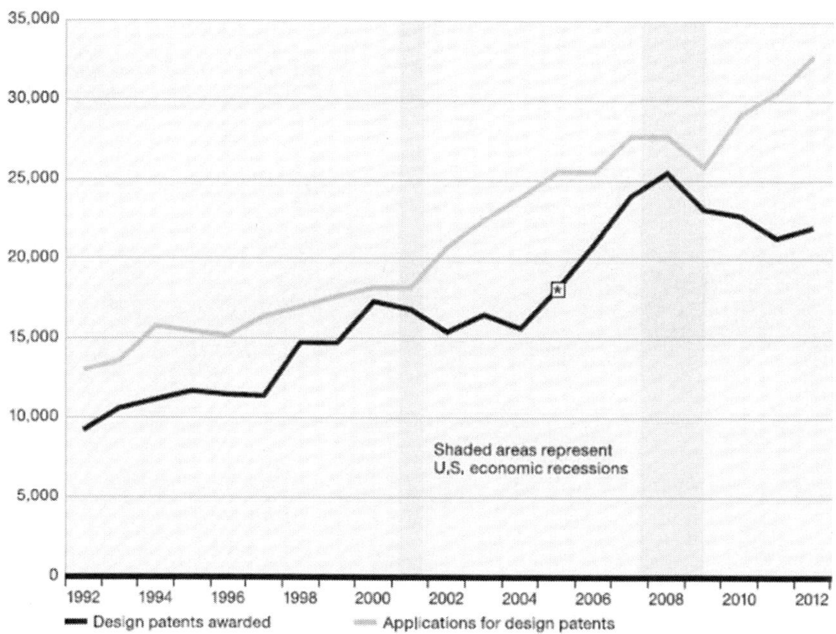

* The USPTO awarded 12,951 design patents in 2005. That count, however, reflects administrative changes, including an aberration resulting from the USPTO's transition from paper to electronic publishing of patents. Consequently, the number of design patents awarded in 2005, shown in the graph above, was calculated by averaging the number awarded in 2004 and 2006, resulting in an estimate of 18,330 design patents awarded in 2005.

Data source: U.S. Patent and Trademark Office, U.S. Department of Commerce.

U.S.-Awarded Design Patents, Grants and Applications, 1992–2012.

Although marked, business cycles are short-run phenomena that can mask long-term trends in design patents. Examination of U.S.-awarded design patents since 1910 reveals two distinctive growth periods: 1910 through the mid-1940s and, in particular, the late 1980s through the present.

To illustrate, the graph below shows a ten-year moving average of U.S.-awarded design patents, per capita. As indicated, design patents witnessed growth between 1910 and the mid-1940s—over this timeframe, U.S.-awarded design patents grew from approximately one design patent per 100,000 population to four.

Remarkable for this period is that unlike utility patents, which plummeted during the 1930s, the number of U.S.-awarded design patents generally grew throughout the Great Depression. Propelling that growth was a milestone in industrial design—"streamlining" introduced by the Burlington *Zephyr* at the 1933–34 Chicago Century of Progress Exhibition (see sidebar on page 45).

However, the mid-1940s through most of the 1960s witnessed a reversal in this trend. In 1958, for example, U.S.-awarded design patents per capita fell to 1.4. The decline during that period reflects, in part, a 1952 revision to patent law that provided short-term protection for ornamental designs of useful articles under a modified copyright approach.

In 1976, however, the U.S. Congress rejected that revision and returned the protection of industrial design to the conventional patent system.[34]

Streamlining: 1933–1934 Chicago Century of Progress Exposition

Manufacturing companies managing to survive the Great Depression of the 1930s faced stiff competition. In order to distinguish their products, companies hired industrial designers who increasingly followed a new design style—streamlining.

Streamlining, made popular by the Burlington *Zephyr*, was introduced at the Century of Progress Exposition at the Chicago World's Fair in 1933 and 1934. It was based on the naturally efficient shapes of fish and birds and incorporated curved and tapered shapes that offered less resistance to air and fluid.

Captivating the American public, streamlined design was evident in objects as varied as cameras, cars, radios, refrigerators, and even the iconic Coca-Cola bottle. Among its chief proponents were Raymond Loewy, who designed the aforementioned Coca-Cola bottle, as well as the Saturn Spaceship and the Greyhound bus and logo, among other products.

Interestingly, Loewy's design studio employed Audrey Moore Hodges, who was the first full-time female auto designer, and who designed the 1947 Studebaker Champion.

Other streamlining designers included Norman Bel Geddes, known for his theatrical and movie sets; Walter Dorwin Teague, associated with exhibition design as well as designs for Ford, Kodak, and Boeing; and Henry Dreyfuss, who famously designed table-top telephones for Bell Laboratories, Hoover vacuum cleaners, and the New York City Hudson locomotive.

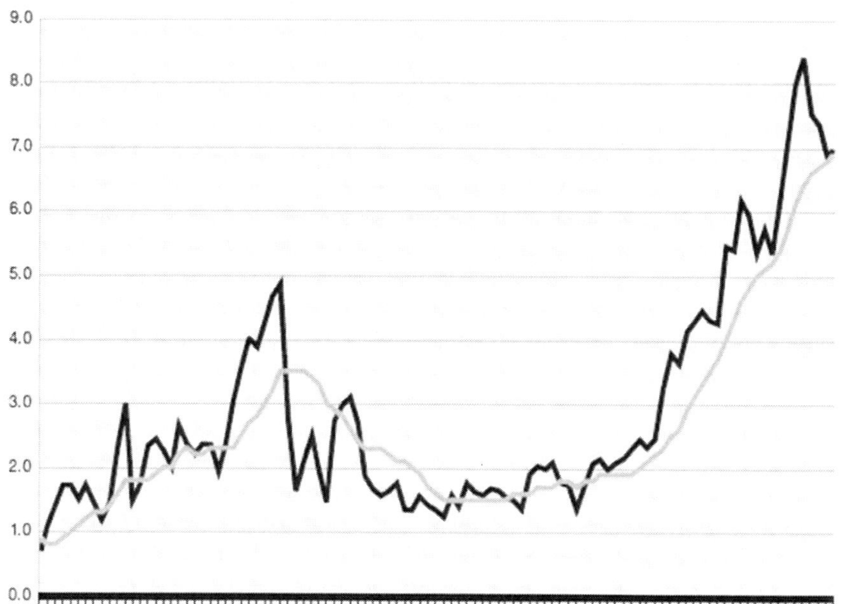

Data source: U.S. Patent and Trademark Office, U.S. Department of Commerce.

History of U.S. Patented Designs per Capita, 1910–2012, by Year Granted.

In the 1980s, consequently, the per-capita concentration of U.S.-awarded design patents rose, heralding a post-industrial era of sustained growth in design and utility patents. In 2008, USPTO-granted design patents per capita reached a high of 8.4. For that matter, the 1980s post-industrial period also witnessed strong growth in *utility* patents, which, per 100,000 population, climbed from 27 in 1980 to nearly 52 in 2008.

Industrial Designers as Inventors

In May 2012, the Brookings Institution convened "The Arts, New Growth Theory, and Economic Development," a National Endowment for the Arts-sponsored symposium to examine "new growth theory" as a tool for assessing the impact of art and culture on the U.S. economy. Presenting at this conference, Alan Marco, currently the USPTO's acting chief economist, portrayed designers as prolific inventors.

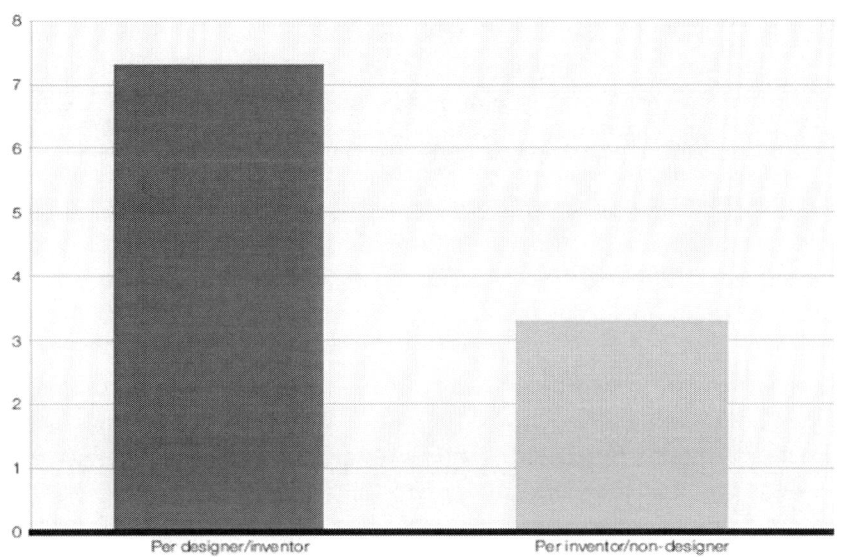

Analysis by Alan Marco, U.S. Patent and Trademark Office, U.S. Department of Commerce.

Average Number of Utility Patents, 1975–2010.

Examining patent filing data between 1975 and 2010, Marco found that, among 136,000 "designers" (those named as inventors on *design* patents), 55,000, or 40 percent, were "designers" and "inventors" (named as inventors on *utility* patents as well).[35] Among the 2.5 million "inventors" during this period, only 2 percent were "inventors" and "designers," i.e., named inventors on both utility and design patents.

Even more telling was the high number of utility patents originating with "designers-inventors." Over the period studied, the mean number of utility patents originating with designers-inventors was 7.3. By contrast, the average number of utility patents naming inventors-designers was 3.3.

CONCLUSION: THE NEA AND INDUSTRIAL DESIGN

At first glance, it may seem as though industrial design has little to do with the National Endowment for the Arts. After all, industrial design clusters around the commercial enterprises of manufacturing and industrial design services.

But that view would be shortsighted. While the Arts Endowment does not award grants to for-profit companies, it does support the schools that train industrial designers and the museums that display and interpret design to the public. In 2013, for example, the agency awarded $50,000 to Atlanta's Robert W. Woodruff Arts Center of the High Museum to support Spark: Innovative Automotive Design, an exhibition that will investigate how visionary automotive designs influenced the automotive industry.

Also, not all industrial design occurs in large, for-profit companies. D-Rev, for example, is a California-based nonprofit product-development company whose mission is to improve the lives of the world's poorest populations. In 2013, D-Rev received a $25,000 NEA grant to develop an interactive online platform for product designers working on social-impact projects.

However, the connection between the NEA and industrial design goes beyond awarding grants. In the fall of 2013, the Arts Endowment, in partnership with the Bureau of Economic Analysis, is scheduled to release the first-ever U.S. satellite account on arts and cultural industries. To fully measure how the arts affect the U.S. economy, the Arts and Cultural Production Satellite Account (ACPSA) will capture traditionally nonprofit arts industries, such as art museums and symphony orchestras, as well as commercial enterprises. The ACPSA, for example, will include a domain labeled "applied arts and design services," which will measure production and employment by specialized design services industries, including industrial design.

Further, the ACPSA, at the time of its release, will be accompanied by an NEA taxonomy of arts and cultural occupations, which will include industrial designers and other design professionals. The ACPSA, and its associated occupations, thus will account for the total economic impact of industrial design, whether in the commercial or nonprofit sector.

As the final section of this report suggests, however, one cannot assign a true value to this industry and its workers without also considering the role of design in product invention. It is to be hoped that future studies, both quantitative and qualitative, will clarify more extensively the link between industrial design and innovation in products and services, but also in processes, systems, and user experiences. A broader frame of inquiry and a mixed-methods research approach would go even further to improve public knowledge about this protean and prolific sector.

APPENDIX

Understanding Industry Classifications

As reported by the U.S. Census Bureau, the North American Industry Classification System (NAICS) is used by federal statistical agencies in classifying business establishments for the collection, tabulation, and presentation of statistical data describing the U.S. economy. The NAICS structure is hierarchical and begins at the two-digit level, which represents the industry sector, and expands to six digits of industry classification detail.

This report, for example, draws on statistics for two industry sectors—manufacturing, which is NAICS 31, 32, and 33, and "professional, scientific, and technical services," which is NAICS 54. Each of these industry sectors comprises industry groups and detailed industries. Professional, scientific, and technical services, for example, includes legal services, accounting and payroll services, computer systems design firms, and advertising and public relations firms, just to name a few.

The professional, scientific, and technical services sector also includes architectural, engineering, and related services (NAICS 5413) and specialized design services (NAICS 5414), two *industry groups* that commonly employ industrial designers. In turn, architectural, engineering, and related services contain architectural services businesses, engineering services firms, and drafting businesses, among others. Specialized design services is the parent group to industrial design firms, fashion design houses, graphic design businesses, and companies selling "other specialized design services" such as jewelry design.

Within the NAICS structure, business establishments are assigned an industry code based on the production or service accounting for the largest share of the establishment's sales or revenue. As discussed in this report, in 2007, 148 industrial design firms (out of a total of 1,637) did $37 million in graphic design business. But these establishments were classified as industrial design firms—the majority of their sales stemmed from product design and other industrial design services.

Illustration of NAICS Structure

NAICS	Hierarchy	Sector, industry group, and industry
54	Industry sector	Professional, scientific, and technical services
5414	*Industry group*	*Specialized design services*
54141	Industry	Interior design services
54142	Industry	Industrial design services
54143	Industry	Graphic design services
54149	Industry	Other specialized design services
5413	*Industry group*	*Architectural, engineering, and related services*
54131	Industry	Architectural services
54132	Industry	Landscape architectural services
54133	Industry	Engineering services
54134	Industry	Drafting services
54135	Industry	Building inspection services
54136	Industry	Geophysical surveying and mapping services
54138	Industry	Testing laboratories

Source: U.S. Census Bureau, U.S. Department of Commerce.

REFERENCES

Andes, S. and Muro, M. (2013). *Jobs Alone Do Note Explain the Importance of Manufacturing*. Advanced Industry Series, The Brookings Institution. Retrieved from http://www.brookings.edu/blogs 04/03-jobs-manufacturing-muro-andes.

Babcock, C. (2012, September 6). "Apple Beats Competition with Design – And Design Patents." *InformationWeek*. Retrieved from http://www.informationweek.com/hardware/handheld/apple-beats-competition 240006830?pgno=1.

Bailey, M.N. and Manyika, J. (2013, January 21). "Is Manufacturing 'Cool' Again?" Brookings Institution. Retrieved from http://www.brookings.edu /research/opinions/2013/01/21-manufacturing-baily-manyika.

Brookings Institution (2012, May 10). The Arts, New Growth Theory, and Economic Development Symposium. Retrieved from http://www.brookings.edu/events/2012/05/10-arts-development#ref-id=20120510_NEA_Panel_1.

DuMont, J. & Janis, M. (2012). *Designing the American Design Patent System.* Retrieved from Social Science Research Network website: http://papers. ssrn. com/sol3/papers.cfm?abstract_id=1862182##.

Employment and Training Administration, U.S. Department of Labor. (2011). *Advanced Manufacturing Industry: Addressing the Workforce Challenges of America's Advanced Manufacturing Workforce*. Retrieved from http://www.doleta. gov/BRG/pdf/Advanced%20Manufacturing%20Report%2011.1.05.pdf.

Industrial Designers Society of America (IDSA). Women in Design Gallery. Retrieved from http://idsa.org/women-design-gallery.

Jones, A. & Vascellaro, J. (2012, July 24). "Apple v. Samsung: The Patent Trial of the Century." *Wal Street Journal*. Retrieved from http://online. wsj.com/article/SB100008723963904432954045775432 21814648592.html.

National Center for Education Statistics (2012). Integrated Postsecondary Education Data System. Retrieved from http://nces.ed.gov/ipeds/.

PBS (Producer). (2006, June 13). "Biography: Industrial Designers and Streamliners." In *American Experience*. Retrieved from http://www.pbs. org/wgbh/americanexperience/features/biography/stream liners-designers/.

Richman, J.H. (1991). "Design Protection and the New Technologies: The United States Experience in a Transitional Perspective." *Baltimore Law Review*, 19. Retrieved from Duke University School of Law website: http://scholarship

The Manufacturing Institute. (2013). Facts About Manufacturing Retrieved fromhttp://www.themanufacturinginstitute.org/Research/Facts-About Man nufacturing/ Facts-2012.aspx.

U.S. Bureau of Labor Statistics (2012). *2012–2013 Occupational Outlook Handbook*. Retrieved from http://www.bls.gov/ooh/.

U.S. Bureau of Labor Statistics (2012). Employment Projections: 2010–2020. Retrieved from http://www.bls.gov/emp/.

U.S. Bureau of Labor Statistics (2013). Labor Force Statistics from the Current Population Survey, Household Data, Annual Averages. Retrieved from http://www. bls.gov/cps/cpsaat12.htm.

U.S. Bureau of Labor Statistics (2013). Occupational Employment Statistics, Overview. Retrieved from http://www.bls.gov/oes/oes_emp.htm.

U.S. Patent and Trademark Office (2013). *Design Patenting by Geographic Region (State and Country), Breakout by Organization*. North Carolina. Retrieved from http://www.uspto.gov/web

U.S. Patent and Trademark Office. (2012). *Design Patent Application Guide*. Retrieved from http://www.uspto.gov/patents

U.S. Patent and Trademark Office (2013). Design Patents Report, Parts A1, A2, B. Retrieved from http://www.uspto.gov/web. htm#PartA1_1.

U.S. Patent and Trademark Office. (2013). U.S. Patent Activity, Calendar Years 1790 to the Present. Retrieved from http://www.uspto.gov/web oeip/taf/h_counts.htm.

Additional Reading

Art Center College of Design. Retrieved from http://www.artcenter.edu.

Car Body Design (2012, June 18). "85 Years of GM Design: The Timeline." Retrieved from http://www.carbodydesign.com/2012/06/gm-design-the-timeline/.

Employment and Training Administration, U.S. Department of Labor. O*Net Online. Retrieved from http://www.onetonline.org/.

Guglielmo, C. (2012, July 31). "Apple Goal with iPhone Was to 'Wow the World,' Not be 'Ripped Off' by Samsung." *Forbes*. Retrieved from http://www.forbes.com/sites/connieguglielmo/2012/07/31/apples-witnesses-take-the-stand-in-patent-trial- with-samsung-live-blog.

Henderson, R. (2012). "Industry Employment and Output Projections to 2020." *Monthly Labor Review*. Retrieved from http://www.bls.gov /opub/mlr /2012/01/ art4full.pdf.

Knoble, J. (1993, June). "Don't Overlook Design Patents." *Design News*.

Langinier, C. & Moschini, G. (2002, February). *The Economics of Patents: An Overview*. Working Paper 02-WP 293. Center for Agricultural and Rural Development, Iowa State University. Retrieved from http://www.card. iastate.edu/publications/DBS/PDFFiles/02wp293.pdf.

Oake, R. (2011, October). "Understanding Functionality in Design Patent Law." *Intelectual Property Today*. Retrieved from http://www.design patentschool.com / assets/Oake_OCT11%20V2.pdf.

Rothwell, J., Lobo, J., Strumsky, D., & Muro, M. (2013). *Patenting Prosperity: Invention and Economic Performance in the United States and its Metropolitan Areas*. Metropolitan Policy Program at Brookings, Brookings Institution. Retrieved from http://www.brookings.edu/research /interactives /2013/metropatenting.

Stanford Program in Law, Science, and Technology (2013, April 5). Design Patents in the Modern World Conference. Retrieved from http://www.

In: U.S. Industrial Design Sector
Editor: Louis H. Nielsen

ISBN: 978-1-62948-371-9

Chapter 2

THE FUTURE OF INDUSTRIAL DESIGN*

Tim Brown

When I was in school back in the 1970s and 80s, industrial design meant creating mass-manufactured products. The prevailing question was *How do you make the perfect cup or chair or sports car?* The challenge was clear: make better, more efficient, more beautiful stuff.

That kind of design still exists, of course, but now the challenge at hand is much more complicated. Why? Because the things we interact with are leaving the simple and physical world and becoming virtual and complex. If industrial design is the interface between the world of technology (defined broadly, to mean anything we design and make) and the world of people, then it has to solve for a whole new level of abstraction. Today, industrial design is as much about designing systems and software and applications as it is about designing objects. We are designing machines, but also the ghosts that live inside them.

A recent IDEO project reflects how much the practice of industrial design has had to stretch in the last couple of decades. For Innova [Schools] (http://www.innovaschools.edu.pe/), we were tasked with reconceiving an entire school system in Peru. The brief included many classic design problems—plans for a campus and classroom—but we were also faced with creating a business model, a curriculum, and digital tool for educators. Moreover, everything we created had to be confluent within a culturally specific system with its own set of requirements. The work still fits my definition of industrial design as a mediation between technology and people,

---

* Posted on the official blog of the National Endowment of the Arts, August 22, 2013.

but because it applies across an entire network of information and activities, it will have far more impact than simply designing a better school desk.

That kind of project is what excites me about being a designer today, and it's why I was surprised that the categories hiring the most industrial designers in the NEA [industrial design] report (automotive, for example) were so traditional. That there wasn't a single service company in most of those tables seemed very 20th-century to me. I'd argue that the opportunity for design—where it's taking the greatest strides—is likely to be outside the categories we'd identify in the past. Perhaps the data simply hasn't caught up with the reality of the situation.

The view from our offices, where we're being asked to design new systems in education and health, looks quite different. Who would have known 15 years ago that Kaiser and the Mayo Clinic would be hiring design teams? Faced with having to support a vast aging population, the health care industry has had to think a lot more like designers.

Not only are we applying our craft to a broader set of industries, we also have unprecedented access to tools and manufacturing techniques that have reunited design with the act of making. And once the making is done, designers can bring their ideas to market in a flash. What's happening on platforms like Kickstarter and Etsy—how effectively they've helped designers with things they're not good at, like raising capital and finding their market—is just jaw dropping. IDEO's extended family has launched many a project on Kickstarter, including a portable kayak called the Oru and "the world's thinnest watch," called CST-01 . Some of today's most exciting startups—Airbnb, Mailbox, Pinterest—have founders who are designers. The crossover between design and entrepreneurship has never been stronger.

So, how do we prepare the next generation of industrial designers for this new abundance? Broadening and deepening educational approaches to design is a good start. Art school was invented by the Bauhaus 100 years ago, and that has been the go-to model ever since. But hybridized programs are starting to pop up, like the MIT media lab and Stanford's d. school (started by IDEO's very own David Kelley). Design is a practice that one learns by doing, not by listening to lectures. We need to capture the minds of more kids, tap into a broader segment of the student population, and get them excited about the contribution they can make.

To summarize William Gibson, *the future is already here—it's just not evenly distributed.* But that moment is coming. It's exciting to see this pivot from designing artifacts to inventing entire systems that we live in and operate. More than ever before, design is transforming how we think and behave and connect with other people—and the world.

INDEX

J

L

M

N

O

P

R

S

T

U

V

W

POLLUTION SCIENCE, TECHNOLOGY AND ABATEMENT

MICROPOLLUTANTS

SOURCES, ECOTOXICOLOGICAL EFFECTS AND CONTROL STRATEGIES

TABITHA N. HOLLOWAY

EDITOR

Library of Congress Cataloging-in-Publication Data

ISBN: 978-1-53612-067-7

Published by Nova Science Publishers, Inc. † New York

CONTENTS

PREFACE

Many organic pollutants, such as, organochlorine pesticides (OCPs), polychlorinated biphenyl (PCBs) and polycyclic aromatic hydrocarbons (PAHs) are detected in the environment at low concentrations. Due to their large-scale production and usage, toxicity, bioaccumulation and persistence in the environment, they can cause harmful effects to organisms and to human health. Atmospheric deposition, industrial and urban activity presents the main sources of environmental pollution. Chapter One covers inorganic pollution by heavy metals, which is less visible and direct than other types of pollutants, but its effects on marine ecosystems and humans are intensive and very extensive. Chapter Two discusses the degradation resistance of Di(2-ethylhexyl) phthalate (DEHP), an artificially synthethized organic compound extensively used as plasticizers for industrial, medical, and domestic purposes, and that is found in wastewater treatment plants. Chapter Three discusses how the pollution risk of soil contamination by the lanthanides and platinum group metals is increasing with the industrialization and urbanization growth in Russia.

Chapter 1 – Many organic pollutants, such as, organochlorine pesticides (OCPs), polychlorinated biphenyl (PCBs) and polycyclic aromatic hydrocarbons (PAHs) are detected in the environment at low concentrations.. Due to their large-scale production and usage, toxicity, bioaccumulation and persistence in the environment, they can cause harmful effects to organisms and to human health. Atmospheric deposition, industrial and urban activity presents the main sources of environmental pollution. Inorganic pollution by heavy metal (HM) is less visible and direct than other types of pollutants, but its effects on marine ecosystems and humans are intensive and very extensive. Organic and inorganic compounds must have a control strategy to detect and control levels of pollution in the environment. Atomic absorption spectrometry (AAS) is being used for the analyses of heavy metal. Analyses of organic pollutants are being performed by gas chromatography (GC) combined with an adequate detector. Electron captor detector (ECD), flame ionization detector (FID) or mass spectrometry (MS) detector was mostly used for the detection of organic pollutants. Many conventions have been signed and legislations have been established to protect human health and the environment from pollution. Examples of OCPs, PAHs, PCBs and heavy metals studies in different matrices in many regions in Tunisia have been presented.

Chapter 2 – Phthalate esters (PAEs) are artificially synthetized organic compounds extensively used as plasticizers for industrial, medical, and domestic purposes. Di(2-ethylhexyl) phthalate (DEHP) is one of the most synthetized PAEs and is considered resistant to the biological degradation due to its long hydrocarbon chain. An indigenous microorganism was isolated from the sludge collected from a local wastewater treatment plant (Macau SAR, China) to remove DEHP from artificially contaminated water. The 16S rRNA gene sequence analysis identified the microbial strain as *Acinetobacter* sp. SN13. Such major experimental parameters as pH (6-9) and temperature (35°C) were further optimized to improve the DEHP biodegradation efficiency. The growth kinetics followed the inhibition model (simulated using Matlab), with half saturation constant (272.3 mg l^{-1}), maximum degradation rate (124.8 mg l^{-1} day^{-1}), and inhibition constant (720.5 mg l^{-1}) estimated for the DEHP degradation, and half saturation constant (137.6 mg l^{-1}), specific growth rate (0.1192 day^{-1}), and inhibition constant (850.3 mg l^{-1}) for the microbial growth on DEHP. Since many environmental sites are contaminated with a mixture of inorganic and organic contaminants, the effect (inhibitory/stimulatory) of some microelements commonly present in wastewater (Fe^{3+} and Mn^{2+}) on DEHP biodegradation was also evaluated. The biodegradation performance of the isolate was improved as the Fe^{3+} concentration increased (100-1,000 $\mu g\ l^{-1}$), while higher Mn^{2+} concentrations (500-1,000 $\mu g\ l^{-1}$) inhibited the DEHP biodegradation. The aerobic biodegradation of phthalates generally occurs in two stages. First, phthalate diesters (PDEs) hydrolyze to phthalate monoesters (PMEs) followed by the PMEs hydrolysis to phthalic acid (PA) and then, the PA mineralization takes place by different mechanisms. For the Gram-negative bacteria like *Acinetobacter* sp., PA is usually further degraded via the dioxygenase-catalyzed pathways to protocatechuate (3,4-dihydroxy-benzoate) through 4,5-dihydroxyphthalate and *cis*-4,5-dihydroxy-4,5-dihydrophthalate. The respective DEHP degradation pathway for *Acinetobacter* sp. SN13 is proposed through the identification of mono-(2-ethylhexyl) phthalate (MEHP), PA, 3-katoadipate, β-carboxy-*cis,cis*-muconic acid, and protocatechuate by LC-MS.

Chapter 3 – Pollution risk of soil contamination by the lanthanides (Ln) and platinum group metals (PGM) increases with industrialization and urbanization growth in Russia. The development of electronic, oil chemistry, metallurgy, medical industries rises the input of Ln-containing wastes into the soil. When considering soil contamination by Ln and PGM, it appears important to point out the way: basically, aerial, although hydrogenic contamination of alluvial soils is observed in some places too. Increasing the number of combustion engine with catalytic exhaust neutralizers, led to air and soil pollution by cancerogenic PGM: rhodium, palladium, platinum. In Russia, road dust analysis is widely used in the study of aerial soil contamination by Ln and PGM. Soils in Russia are contaminated by Ln and PGM mostly by aerial way near steel plants, thermal power coal-fired plants, as well as in large cities with tight traffic. Alluvial soils along rivers are contaminated hydrogenically by untreated industrial wastewater or by leachating from dumps, where waste is stockpiled after non-ferrous metals ores enrichment. In hydrogenically contaminated alluvial soils micropollutants are strongly accumulated in Fe-

Mn-nodules, due to this eliminating from the geochemical cycle. The peat soils in the Western Siberia are contaminated by the lanthanides in the sites of oil spill.

In: Micropollutants
Editor: Tabitha N. Holloway
ISBN: 978-1-53612-067-7

Chapter 1

THE DISTRIBUTION OF ORGANIC AND INORGANIC POLLUTANTS IN MARINE ENVIRONMENTS

Mouna Necibi[1,2,*] and Nadia Mzoughi[3,†]
[1]Département de chimie, Faculté des Sciences de Tunis, Université de Tunis EL Manar, Tunisie
[2]Laboratoire Milieu Marin, Institut National des Sciences et Technologies de la Mer, La Goulette, Tunisie
[3]Laboratoire des Sciences et Technologies de l'Environnement, Institut Supérieur des Sciences et Technologies de l'Environnement, Borj Cédria, Université de Carthage, Tunisie

ABSTRACT

Many organic pollutants, such as, organochlorine pesticides (OCPs), polychlorinated biphenyl (PCBs) and polycyclic aromatic hydrocarbons (PAHs) are detected in the environment at low concentrations. Due to

* Corresponding Author Tel: (216) 71 872 600, Fax: (216) 71 871 666. Tel. +216 71 730 420; Fax: +216 71 732 622; Email: mouna23necibi@gmail.com.
† Corresponding Author Email: nadia.mzoughi@instm.rnrt.tn. Tel/Fax: +216 79325333.

their large-scale production and usage, toxicity, bioaccumulation and persistence in the environment, they can cause harmful effects to organisms and to human health. Atmospheric deposition, industrial and urban activity presents the main sources of environmental pollution. Inorganic pollution by heavy metal (HM) is less visible and direct than other types of pollutants, but its effects on marine ecosystems and humans are intensive and very extensive. Organic and inorganic compounds must have a control strategy to detect and control levels of pollution in the environment. Atomic absorption spectrometry (AAS) is being used for the analyses of heavy metal. Analyses of organic pollutants are being performed by gas chromatography (GC) combined with an adequate detector. Electron captor detector (ECD), flame ionization detector (FID) or mass spectrometry (MS) detector was mostly used for the detection of organic pollutants. Many conventions have been signed and legislations have been established to protect human health and the environment from pollution. Examples of OCPs, PAHs, PCBs and heavy metals studies in different matrices in many regions in Tunisia have been presented.

Keywords: micropollutants, marine environment, toxicology effects, control strategy

INTRODUCTION

Aquatic ecosystems are important reservoirs of pollutants, some of which are toxic and their release into the environment may threaten the balance of aquatic ecosystems and human health. The marine environment pollution sources are various and affect many countries, including Tunisia. Among these pollutants, there are industrial sources, agricultural, domestic and natural sources, which can be direct or indirect, individual or diffuse. The main types of pollutants that reach the aquatic environment and their origins are either organic compounds represented mainly by polycyclic aromatic hydrocarbons, pesticides, drugs, dioxins, chlorophenols, phthalates, or inorganic compounds which can be represented by metal compounds, nitrogenous materials and phosphorus. The organochlorine pesticides (OCPs) and polychlorinated biphenyls (PCBs) are among the most important persistent organic pollutants (POPs) and they are well known by their chronic toxicity, persistence, and bioaccumulation.

Polycyclic aromatic hydrocarbons (PAHs) are one of the most widespread organic chemicals originated mainly from incomplete combustion of organic materials and accidental spillage of fossil fuels. Due to their carcinogenic or mutagenic effects to both terrestrial and aquatic organisms, the transport and fate of PAH in coastal areas have received much attention. Because of the different intrinsic physicochemical properties of individual PAH, such as solubility, vapor pressure and lipophilicity, they tend to interact to different extents with sediments, suspended particulate matter (SPM), water and biota. PAHs are subject to various transformation processes including, chemical transformation, biodegradation and photochemical degradation.

Heavy metals are natural constituents of the marine environment. These metals always function in combination with organic molecules, usually proteins. Metals occur normally at low concentrations yet they have the capacity of making significant considerable biological effects even at such levels. All metals are toxic at levels higher than threshold limit. Silver (Ag), mercury (Hg), copper (Cu), cadmium (Cd) and lead (Pb) are particularly toxic. Metal pollution of the marine environment is less visible and direct than other types of marine pollution, but its effects on marine ecosystems and humans are intensive and very extensive. Despite the prohibition of the use of some of these pollutants in North America and Europe for over 40 years and in Tunisia since the 80s, they continue to pose problems because of their persistence in the environment. The sediments may therefore represent an important source of contaminants for communities of organisms that live there. Regulatory measures (prohibition or limitation of the use of certain chemical compounds) are also applied in order to protect humans and their environment.

POLYCHLORINATED BIPHENYLS (PCBS)

Definition

Due to their very low biodegradability in the environment and their tendency of bioaccumulation and biomagnification, they are subject to international restrictions on usage and emissions (Bergman et al. 2012). PCBs are a group of some 209 individual chemical compounds, produced in various industrial mixtures by introducing elementary chlorine in various degrees into biphenyl (see Table 1 and 2). Commercial use of PCBs began in 1929, and production is believed to have finally ceased in the mid-1980s. Figure 1 present chemical structure of polychlorinated biphenyls. Polychlorinated biphenyls PCBs belong to the so called persistent organic pollutants (POPs).

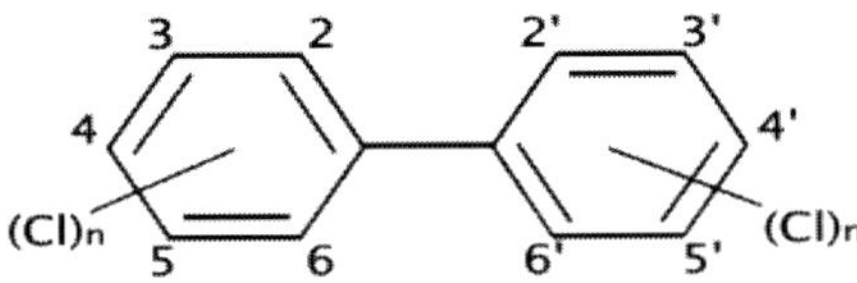

Figure 1. Chemical structure of polychlorinated biphenyls.

Table 1. Classification of different PCBs congeners (Ballschmiter et al. 1980)

Chlorinated Substitutions	Formulas	Number of isomers
Monochlorobiphenyls	$C_{12}H_9Cl$	3
Dichlorobiphenyls	$C_{12}H_8C_{12}$	12
Trichlorobiphenyls	$C_{12}H_7C_{13}$	24
Tétrachlorobiphenyls	$C_{12}H_6C_{14}$	42
Pentachlorobiphenyls	$C_{12}H_5C_{15}$	46
Hexachlorobiphenyls	$C_{12}H_4C_{16}$	42
Heptachlorobiphenyls	$C_{12}H_3C_{17}$	24
Octachlorobiphenyls	$C_{12}H_2C_{18}$	12
Nonachlorobiphenyls	$C_{12}HC_{19}$	3
Décachlorobiphenyls	$C_{12}Cl_{10}$	1

Table 2. Physicochemical data of PCBs for each group of isomers (Robertson et al. 2001)

Congeners Group	Molecular weight (g mol^{-1})	Vapor Pressure (Pa) at 25°C	Solubility at 25°C (g m^{-3})	Octanol/ water partition coefficient log Kow	Approximate evaporation rate at 25°C (g/(m2/h))
Mono-chlorobiphenyls	188.7	0.9-2.5	1.21-5.5	4.3–4.6	0.25
Dichloro biphenyls	223.1	0.008-0.6	0.06-2	4.9–5.3	0.065
Trichloro-biphenyls	257.5	0.003-0.22	0.015-0.4	5.5–5.9	0.017
Tetra-chlorobiphenyls	292	0.002	0.0043-0.01	5.6–6.5	$4.2\ 10^{-3}$
Penta-chlorobiphenyls	326.4	0.0023-0.051	0.004-0.02	6.2–6.5	10^{-3}
Hexa-chlorobiphenyls	360.9	0.0007-0.012	0.0004-0.0007	6.7–7.3	$2.5\ 10^{-4}$
Hepta-chlorobiphenyls	395.3	$1.3.10^{-4}$	0.000045	6.7–7	$6.2\ 10^{-5}$
Octa-chlorobiphenyls	429.8	$2.8.10^{-5}$	0.0002-0.0003	7.1	$1.5\ 10^{-5}$
Nona-chlorobiphenyls	464.2	$6.3.10^{-6}$	0.00018-0.0012	7.2–8.16	$3.5\ 10^{-6}$
Déca-chlorobiphenyls	498.7	$1.4.10^{-6}$	0.000001	8.26	$8.5\ 10^{-7}$

Source of PCB in the Environment

Emerging sources of PCBs include:

- Increasing demand for electronic equipment, increasing generation of e-waste, and subsequent illegal recycling and treatment of electrical equipment and waste incineration (Breivik et al. 2011).
- Disassembling of transformers with leaching of oil into the environment and dispose of unsalvageable parts to landfill and local vicinity.
- Open burning for recovery of valuable metals, plastic peeling and melting, melting of circuit boards over open fires, and metal extraction in acid baths (Lau et al. 2012).
- Ship wreckage.

Table 3. PCBs applications

Closed applications	Electrical transformers (insulation and cooling liquid) Electrical capacitors (dielectricum), Power factor capacitors in electrical distribution systems, Lighting ballasts, Motor start capacitors in refrigerators, heating systems, air, conditioners, hair dryers, water wheel motors, washing machines, clothes dryers, ventilating fans etc. Capacitors in electronic equipment including TV, microwave ovens, Electrical motors, Electric magnets.
Partially closed applications	PCB-containing oil, Heat transfer fluid (e.g., oil radiator), Hydraulic fluid (e.g., in mines), Vacuum pumps, Switches, Voltage regulators Liquid filled electrical cables, Liquid filled circuit breakers.
Open applications	Lubricants, Immersion oils for microscopes, Brake linings, Cutting oils, Lubricating oils, Adhesives, Special adhesives, Adhesives for waterproof wall coatings, Casting Waxes, Surface Coatings, Paints, Surface treatment for textiles, Carbonless copy paper, Flame retardants (on ceiling tiles, furniture, walls), Dust control (dust binders, asphalt, natural gas pipelines), Inks, Dyes, Printing inks, Plasticizers, Gasket sealers, Sealants in joints in buildings, PVC, Rubber seals (around vents, doors, windows), Insulating materials, Pesticides (additives).

Application of PCBs

They were extensively used between the 1950s and 70s for a broad range of applications as presents in Table 3. They are used as coolants and lubricants in transformers, generators, and capacitors contained in electrical and electronic products, as well as hydraulic and heat exchange fluids because of their general chemical inertness; insulating capacity, heat stability, and low burning capacity (Gioia et al. 2014).

Health Effects of PCBs

The relationship between exposure to PCBs and human health effects is reflective of the large variation in human exposure to the many different congeners and contaminants present in PCB formulations and to combustion by-products of PCBs. The evidence suggests that exposure to

PCBs is associated with increases in the risk of cancers of the digestive system, notably the liver, and malignant melanoma. PCB exposure is also associated with reproductive deficiencies, such as reduced growth rates, retarded development, and neurological effects (although some neurological deficiencies at early ages may disappear later during childhood). PCB exposure cause immunological changes, manifested as increased infection rates, and changes in circulating lymphocyte populations; and dermatological changes, including chloracne and pigmentation disturbances of skin, nails, and gingival, as well as nail deformation after exposure to highly chlorinated congeners (Meador et al. 1996).

Toxicological Effects on Aquatic Organisms

Numerous studies on the effects of PCBs and industrial mixtures (Aroclor) on Fish were carried out and demonstrated changes in hormone levels, Thyroid disorders (T4 and T3), and therefore endocrine disruption, especially in The European bar (Dicentrarchus labrax), the brown bullhead (Ameiurus nebulosus), the Medaka (Oryzias latipes) and juvenile rainbow trout (Oncorhynchus mykiss) (Yum et al. 2010). Studies of juveniles or adults fishes have shown that PCBs. Also have an effect on growth even if the results remain mixed. For example, Exposure to PCBs reduces growth of eel (Anguilla anguilla) and larvae of Micropogonias undulates (Van Ginneken et al. 2009). Other work also indicates that sediments or food containing PCBs Can affect the rate of fertilization of fish.

Legislation and Convention for PCB

The use of PCBs in open applications such as printing inks and adhesives was banned in the European Community in 1976. Use of PCBs as a raw material or chemical intermediate has been banned in the EU since 1985. In 1996 the directive was replaced by Directive 96/59/EC, which set a deadline of 2010 for completely phasing out or decontaminating any equipment containing PCBs. However, the united nation environment program (UNEP) and the global treaty (Stockholm Convention) on Persistent Organic Pollutants (May 2001) stipulates that the use of equipment shall be eliminated by 2025. The Stockholm Convention came into force in May 2004 after initial ratification by 128 states. As of December 2008, 168 states were a party to the convention (APEK 2005). The Basel Convention "May 1992" adopted technical guidelines for PCB handling, disposal and destruction.

Example of Study: Distributions of Polychlorinated Biphenyl in Surface Water from the Bizerte Lagoon, Tunisia (Necibi et al. 2015)

The contamination of the marine environment by organic pollutants is a matter of great concern. Bizerte Lagoon is the second largest lagoon in Tunisia, and is known for different industrial and fishing activities. Extraction of water samples was performed by liquid-liquid extraction with hexane. Gas chromatography with ^{63}Ni electron capture detector (GC-ECD) was used to perform qualitative and quantitative determinations.

Total PCBs (∑PCB) concentrations varied between 3 and 10.4 ng L^{-1} (Table 4). Study areas with sampling locations and different types of industrial areas (A, B, C and D) in the Bizerte Lagoon were presented in Figure 2. The highest ∑PCB were observed for station S14 and S9 with a concentration of 10.4 ng L^{-1} in the two stations and the lowest levels of ∑PCB were found at station S3 (3 ng L^{-1}) and S13 (2,4 ng L^{-1}). The concentration of PCB 101 in the surface water ranged from ND to 2.8 ng L^{-1}. PCB 28, PCB 180 and PCB 52 were the most predominant congeners, with concentrations varying from 0.2 to 1.4 ng L^{-1}, ND to 6.7 ng L^{-1} and 0.1 to 3.5 ng L^{-1} respectively. The highest concentration of PCB 153 was 3.3 ngL^{-1} obtained for S2. The levels of PCB 138 varied between 0.5 and 5.7 ng L^{-1}. Except station 12, PCB 209 is under detection limits in all sample sites. PCB 180 was detected in the majority stations except S1 an S13. Stations 9 and 14 appear as the most contaminated with ∑PCBs concentrations around 10.4 ng L^{-1}. Due to the location of these two stations, the contamination can be explained by the anthropogenic sources (industrial areas C and D).

Considering the water flow until the lagoon, it is interesting to notice the PCBs diffusion to station 10 and 11, where a concentration gradient appears. Generally, potential sources of PCBs in marine environment are due to the waste of electrical transformers, oil spillage, and any historical use of PCB (Wurl et al. 2005). Low PCBs concentrations found are probably due to their low solubility in water and high octanol–water partition coefficients, once arrived to the marine environment, PCBs can be associated with suspended particulate matter and ultimate by accumulated in bottom sediments (Yang et al. 2009).

ORGANOCHLORINE PESTICIDES (OCPs)

Definition

Organochlorine pesticides (OCPs) comprise several groups of chemicals, but tend to share certain characteristics and structural features. They typically have an aliphatic or aromatic cyclical structure, which is heavily substituted with chlorines. As a result, most OCPs are sparingly soluble and semivolatile. The OCPs can remain unchanged for a long time in the environment. Because of this persistence, phase distribution and transport processes tend to play a larger role in controlling their environmental fate and bioaccumulation behavior than for other more readily degradable substances (Shen et al. 2005). Table 5 and 6 presented different physicochemical properties of the organochlorine pesticides.

Source and Use of Organochlorine Pesticides

Aldrin: Popular pesticide for crops like corn and cotton. It was applied to the soil to control root worms, beetles, and termites. It was used as mothproofing in manufacturing processes and dipping roots and tops of nonfood plants.

Table 4. Limit of detections and concentrations of organochlorine pesticides and polychlorinated biphenyls ($ng\ L^{-1}$) in surface water collected from Bizerte Lagoon

Compounds	Limit of detection	S1	S2	S3	S4	S5	S6	S7	S8	S9	S10	S11	S12	S13	S14
PCB 28	0.018	0.9	1.4	0.6	1.2	0.2	0.4	0.8	1.2	1.1	0.7	0.9	1.0	0.4	1.3
PCB 52	0.018	3.5	0.1	0.8	0.6	0.5	2.0	3.5	3.1	1.8	0.4	1.8	0.4	1.8	0.8
PCB 101	0.014	ND	ND	ND	ND	0.8	0.9	ND	2.5	ND	ND	ND	2.0	0.2	2.8
PCB 138	0.014	ND	ND	ND	ND	1.5	0.5	ND	1.4	ND	5.7	ND	ND	ND	1.6
PCB 153	0.014	ND	3.3	0.8	0.6	2.8	2.1	1.6	0.2	0.8	0.2	1.3	ND	ND	1.5
PCB 180	0.014	ND	2.3	0.8	2.5	2.5	1.2	0.5	1.9	6.7	1.5	5.1	0.8	ND	2.4
PCB 209	0.014	ND	ND	ND	ND	ND	ND	ND	ND	ND	ND	ND	1,1	ND	ND
∑PCBs	-	4.4	7.1	3	4.9	8.3	7.1	6.4	10.3	10.4	8.5	9.1	5.3	2.4	10.4

Table 5. Physicochemical data of the organochlorine pesticides (Zitko et al.2002)

Compounds	classification	Name	Formulas	Chemical structure
HCB	Fongicide	3,4,5,6,9,9,-hexachloro-1a,2,2a,3,6,6a,7,7a-octahydro-2,7:3,6-dimethanonaphth [2,3-b] oxirene.	C_6Cl_6	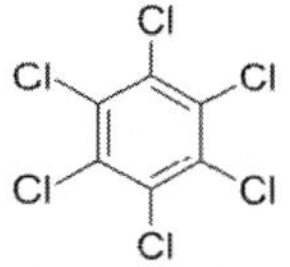
Lindane	Insecticide	1α,2α,3β,4α,5α,6α)1,2,3,4,5,6 hexachlorocyclohexane	$C_6H_6Cl_6$	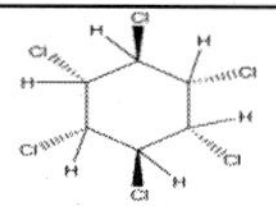
Heptachlor	Insecticide	1,4,5,6,7,8,8-heptachloro-3a,4,7,7a-tetrahydro-4,7-methanol-1H-indene.	$C_{10}H_5C_7$	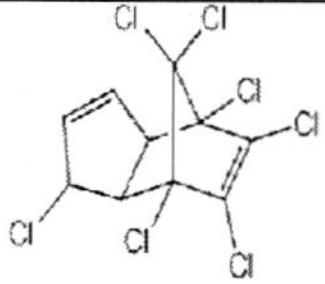
Aldrin	Termiticide	1,2,3,4,10,10-Hexachloro-1,4,4a,5,8,8a-hexahydro-1,4:5,8-dimethanonaphthalene.	$C_{12}H_8C_{16}$	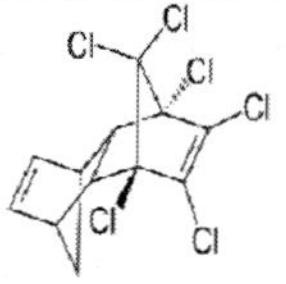
Dieldrin	Insecticide, Termiticide	3,4,5,6,9,9-Hexachloro-1a,2,2a,3,6,6a,7,7a-octahydro-2,7:3,6-dimetanonapth[2,3-b]oxirene	$C_{12}H_8C_{16}O$	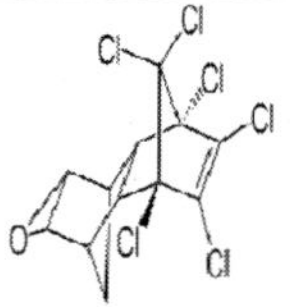

Table 5. (Continued)

Compounds	classification	Name	Formulas	Chemical structure
Endrin	Insecticide	3,4,5,6,9,9,-Hexachloro-1a,2,2a,3,6,6a,7,7a-octahydro-2,7:3,6-dimethanonaphth[2,3-b]oxirene.	$C_{12}H_8C_{16}O$	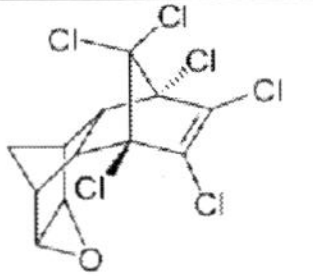
pp' DDT	Insecticide	1,1,1-trichloro-2,2-bis(4-chlorophényl) étane	$C_{14}H_9C_{15}$	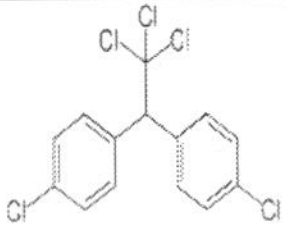
pp' DDD	Insecticide	1,1,-dichloro-2,2-bis(4chlorophényl) éthane	$C_{14}H_{10}C_{14}$	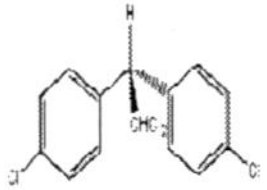
pp' DDE	insecticide	1,1-dichlro-2,2-bis(4-chlorophényl) éthyléne	$C_{14}H_8C_{14}$	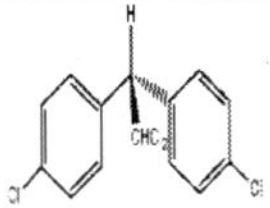

Table 6. Physicochemical properties of the organochlorine pesticides (Gao et al. 2008)

compounds	Weight Molecular (g mol^{-1})	Solubility (mg L^{-1})	Cste H (Pa.m3/mol)	Log Kow	K oc (l kg^{-1})
α-endosulfan	406.9	0.15 – 0.32 at22°C	1.01 - 10,2	3.83 - 4,74	3550 - $2.13.10^4$
β endosulfan	406.9	0.33 at 22°C	1.94	3.52	8600 - $1.39.10^4$
α-HCH	290.83	1.59 – 2.0 at 20°C	1.06 à 20°C	3.77	1760 - 3800
β-HCH	290.83	0.24 – 0.32 at 20°C	0.074 à 20°C	3.85	2140 - 3800
γ-HCH (lindane)	290.83	6.8 – 7.0 at 20°C	0.35	3.69	1350 - 4800
DDT	354.49	0.025 at 25°C	0.84	6.91	$1.51.10^5$
DDE	318.03	0.12 at 25°C	2.13	6.51	$5.01.10^4$
DDD	320.05	0.09 at 25°C	0.41	6.02	$1.51.10^5$
Aldrine	364.91	0.011 at 20°C	4.96	6.50	$4.68.10^7$
Dieldrine	380.91	0.11 at 20°C	0.53	6.20	$4.68.10^6$
Endrine	380.9	0.2 at 25°C	0.041 – 0.055	5.34 – 5.6	3.40.104 - $1.57.10^5$
Isodrine	-	-	-	-	-
HCB	284.79	5.10^{-3} at 25°C	131	5.5	363 - $3.40.10^4$

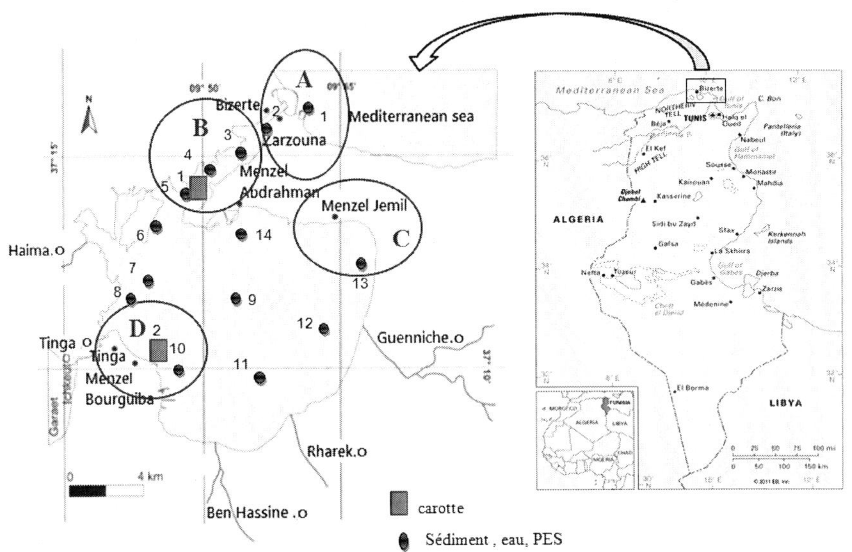

Figure 2. Study areas with sampling locations and different types of industrial areas (A, B, C and D) in the Bizerte Lagoon.

DDT: Used on agricultural crops to control disease carrying insects in communities. Used on animals and humans to control ticks and lice.

Dieldrin: Used on crops like corn, cotton, citrus, soil and seed treatment. Used to control disease vectors, like mosquitoes. Used for sheep dip used prior to shearing and also for wood preservative. Used for mothproofing in manufacturing, including wool products. Used as a dip for the roots and tops of nonfood plants. Used for termite control.

Endosulfan: used on a wide variety of food crops (tea, fruits, vegetables, and grains), ornamental plants, trees and wood preservative.

Endrin: Used for pest control, including termites, mice and army worms. Used as insecticide on crops like cotton, citrus, potatoes, wheat, apples, and flowers, and bark treatment of ash and Hackberry tree.

Heptachlor: Used for public health purposes to control disease vectors, like mosquitoes. Insecticide for homes and buildings, lawns, gardens, cotton, and some food crops, like corn. Used for control of cotton boll weevil. Seed coating, except for termiticide and fire ant control.

Hexachlorobenzene: used as a shield protector (especially wheat seeds) and for a variety of non-pesticide all industrial purposes.

Lindane: used on food crops and forests. Used to control lice and scabies in live stock and humans. It has therapeutic uses included pediculicide (to kill lice), scabicide, and veterinary. It is well known as an ectoparasiticide (to kill external parasites, like fleas) in pharmaceutical products for the treatment of head lice and scabies. (Ritter et al. 1995).

Health Effect of Organochlorine Pesticides

The health effects of organochlorine pesticide exposure depend on the specific pesticide, the level of exposure, the timing of exposure and the individual. Different pesticides result in a range of health symptoms. Numerous studies have linked organochlorine pesticide exposures with cancers and other health effects. Exposure to DDT has been linked to pancreatic cancer and non Hodgkin's lymphoma (Garabrant et al. 1992). Many other organochlorine pesticides, such as Mirex, chlordane and toxaphene, are known to be carcinogenic as well. The organochlorine pesticide exposure is associated with Neuro-developmental health effects in humans. Exposure to organochlorine pesticides has been linked to decreased psychomotor function and mental function, including memory, attention, and verbal skills in children (Jurewicz et al. 2008). Maternal exposure to beta-HCH, a byproduct of lindane manufacture, is associated with preterm births. Dieldrin exposure was associated with decreased T4 levels. Organochlorine chemicals, including DDT, endosulfan and lindane, are known to produce anti-thyroid effects. Thyroid hormones are critical for normal growth and development in fetuses, infants, and small children (Palanzaa et al. 1999). Thyroid deficiencies during pregnancy and postpartum are known to cause altered development, retardation, decreased intellectual capacity, psychomotor delays, and deafness. Neurodegenerative diseases such as Parkinson's disease and Alzheimer's disease are more common in people with general pesticide exposures, including organochlorine pesticide exposure. Classification of the carcinogenicity of the organochlorine pesticides studied was presented in Table 7.

Table7. Classification of the carcinogenicity of the organochlorine pesticides

Compounds	UE	IARC	US-EPA
α-endosulfan	-	-	-
β endosulfan	-	2B	B2
α-HCH	-	2B	C
β-HCH	-	2B	-
γ-HCH (lindane)	-	2B	B2
DDT	3	2B	B2
DDE	-	-	B2
DDD	-	-	B2
Aldrin	3	3	B2
Dieldrin	3	3	B2
Endrin	-	3	D
Isodrin	-	-	-
HCB	2	2B	B2

2 (UE): Assimilated carcinogen,
3 (UE): Substance of concern,
2B (IARC): possibly carcinogenic
3 (IARC): not classified,
B2 (US-EPA): Probably carcinogenic,
C (US-EPA): possibly carcinogenic
D (US-EPA): not classified,

Table 8. Concentration limits of OCPs in water matrix (ng L^{-1})

Compounds	Classification	WHO	Australia guidelines	European Directive
HCB	Ia[a]	1000	-	100
Lindane	II[b]	2000	50	100
Heptachlor	O[c]	30	50	30
Aldrin	O[c]	30	10	30
Dieldrin	O[c]	30	10	30
Endrin	O[c]	-	-	100
pp' DDT	II[b]	2000	60	100
pp' DDD	II[b]	2000	60	100
pp' DDE	II[b]	2000	60	100

a: Extremely hazardous, b: Moderately hazardous; c: Obsolete as pesticide, not classified

Conventions and Legislation for Organochlorine Pesticide

The public interest concerning pesticide residues in consumer products has increased steadily in the last twenty-five years (Haib et al. 2003) and led authorities, legislation to establish strict regulations and monitoring of the quality of products.

These actions are carried out in order to avoid risks to the consumer, but also to regulate the international market (Nunez et al. 2005). Since the early 80s, the European Union has gradually framed pesticide use by various regulations aimed at reducing the impacts proven environmental and risks to human health. The Directive 91/414/EC adopted in 1991 by the European Council aims to assess the health and environmental risks of pesticides used in agriculture in order to optimize the protection of humans and environments. Among the most prominent examples include the withdrawal in 1998 of lindane used in soil treatment against subterranean pests, and recently the withdrawal of triazines, the most widely used herbicides, due to the presence of residues in groundwater and superficial superior to European standards. In Tunisia during the 1940s and 1980s, the use of DDT was widespread, but was banned in early 1984. Tunisia became a party to the Stockholm Convention on 17 June 2004. The chemicals restricted by the Stockholm Convention are the pesticides aldrin, chlordane, dieldrin, endrin, heptachlor, hexachlorobenzene (HCB), Mirex and toxaphene, as well as the industrial chemical PCBs (APEK. 2005). The European Directive 98/83/EC regulated the standard limits of pesticides in water reserved for human consumption. The maximum allowed concentration of each pesticide in water was set at 0.1 ng $mL^{-1}$for individual pesticides and at 0.5 ng mL^{-1} for the total amount of pesticides. The maximum individual concentration for aldrin, dieldrin, and heptachlor must not exceed 0.03 ng mL^{-1} (EC 1998). Table 8 presented concentration limits of OCPs in water matrix (ngL^{-1}) according to WHO, Australia guidelines and the European directive.

Example of Study: Distributions of Organochlorine Pesticides in Surface Water from the Bizerte Lagoon, Tunisia (Necibi Et al. 2015)

Levels of OCPs compounds in surface water samples are represented in Table 9. Study areas with sampling locations and different types of industrial areas (A, B, C and D) in the Bizerte Lagoon were presented in figure 2. Concentration of total OCPs ranged from 0.42 and 14.92 ng L^{-1}. The less contaminated stations in the lagoon were S5, S8 and S13 with ∑OCPs concentration 0.42, 0.45 and 0.47 ng L^{-1} respectively. It can probably be attributed to water flow, especially for S8 situated in the mouth of Tinja River connecting the lagoon to lac Ichkeul. The most polluted stations were 6, 10 and 12 with ∑OCPs concentrations 9.1, 14.92 and 10.19 ng L^{-1}, respectively. Due to the location of these three sampling points the contamination can be attributed to activities from area D and possibly to drain water from agriculture area near to Guenniche and Ben Hassine rivers. The pollution for the rest of the stations is medium with total pesticide concentration ranging between 0.51 ng L^{-1} for station 1 and 6.67 ng L^{-1} for station 7. Because of their lipophilic and hydrophobic nature OCPs tend to accumulate more in the organic phase of sediment and organisms (Adeyemi et al. 2011). At the same time, several physical and chemical processes that include volatilization, migration, photolysis, hydrolysis, aquatic animal metabolism, absorption, also occur (Liu et al. 2013). Sediments represent a source from which OCPs are reintroduced into the water column, and therefore contribute to measurable concentration in the water and biota of many surface water systems (Kuranchie-Mensah et al. 2012).

Polycyclic Aromatic Hydrocarbons (PAHs)

Definition

PAHs represent a family of more than 100 organic molecules comprising at least two aromatic cycles. They are divided into two categories: low-weight molecular compounds (less than 4 aromatic cycles) and compounds of high molecular weight (4 cycles or more). PAHs represent a complex mixture of compounds originating from the incomplete combustion of organic matter. PAHs are hydrophobic molecules. They are classified as persistent organic pollutants (POPs) by the United Nations Environmental Program (Mastral et al. 2000). Figure 3 presented chemical structures of PAH.

Naphtalene Acenaphtylene Acenaphtene Fluorene

Anthracene Fluoranthene Pyrene Benz(a)anthracene

Benzo(b)fluoranthene Benzo(k)fluoranthene Benzo(j)fluoranthene Benzo(e)pyrene

Benzo(a)pyrene Perylene Dibenz(a,h)anthracene Phenanthrene

Dibenz(a,c)anthracene Benzo(g,h,i)perylene Indeno(1,2,3cd)pyrene Chrysene

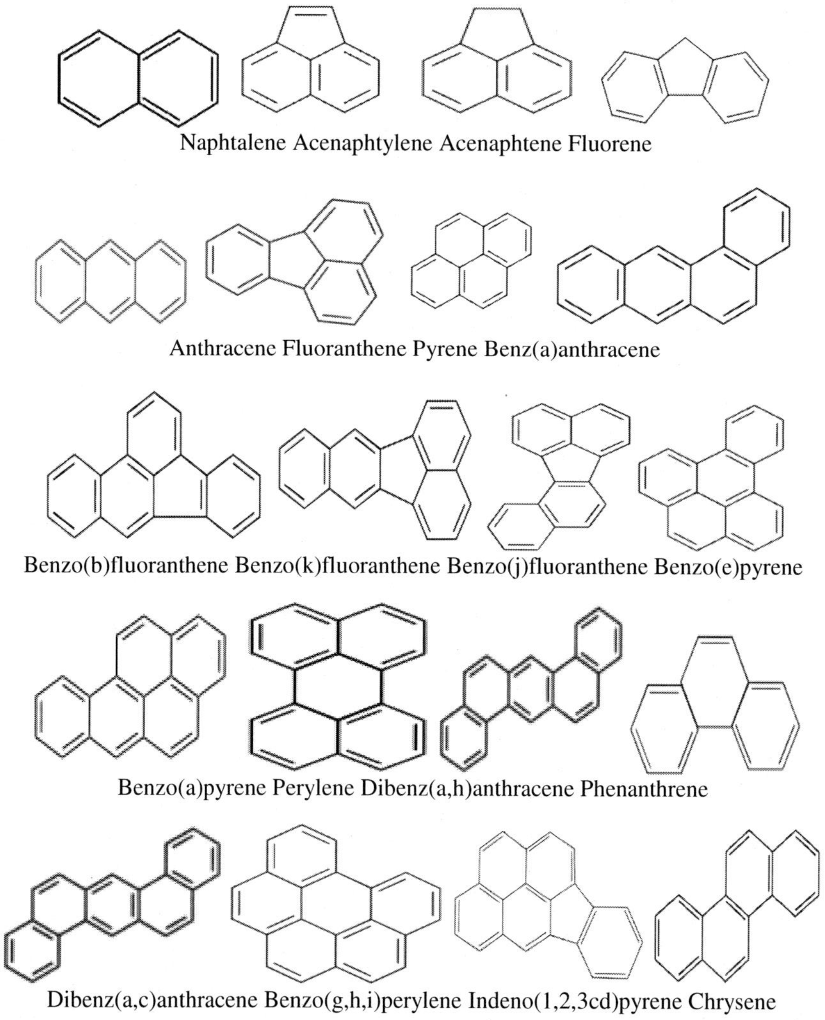

Figure 3. Chemical structures of PAH.

Table 9. Limit of detections and concentrations of organochlorine pesticides and polychlorinated biphenyls ($ng\ L^{-1}$) in surface water collected from Bizerte Lagoon

Compounds	Limit of detection	S1	S2	S3	S4	S5	S6	S7	S8	S9	S10	S11	S12	S13	S14
HCB	0.018	0.19	0.59	2.18	0.34	0.13	0.34	0.24	0.10	1.27	0.48	0.08	0.12	0.12	0.63
Heptacolor	0.018	ND	ND	ND	ND	ND	ND	ND	ND	ND	5,35	4.58	7.83	ND	ND
Lindane	0.006	0.02	ND	ND	0.04	0.05	ND	ND	ND	ND	ND	0.06	ND	0.02	0.03
Aldrin	0.016	0.3	0.54	ND	ND	ND	ND	ND	ND	0.76	0.60	0.29	ND	0.18	2.02
Dieldrin	0.017	ND	ND	ND	ND	0.06	ND	ND	ND	ND	ND	ND	ND	ND	ND
Endrin	0.017	ND	ND	ND	0.17	0.18	0.25	0.21	0.35	0.22	1.51	ND	0.23	0.15	0.26
pp'DD	0.006	ND	ND	ND	ND	ND	2.26	2.39	ND	ND	1.23	ND	ND	ND	1.10
pp'DDE	0.006	ND	3.67	ND	ND	ND	6.25	ND	ND	ND	ND	ND	ND	ND	ND
pp'DDT	0.006	ND	ND	ND	ND	ND	ND	3.83	ND	ND	5.75	ND	2.01	ND	1.10
∑DDTs	-	ND	3.67	ND	ND	ND	8.51	6.22	ND	ND	6.98	ND	2.01	ND	2.20
∑OCPs	-	0.51	4.8	2.18	0.55	0.42	9.1	6.67	0.45	2.25	14.92	5.01	10.19	0.47	5.14

Table 10. Physical and chemical characteristics of PAHs (Jiang et al. 2012)

PAH	Ab	Molecular weight (g mol^{-1})	Vapor Pressure (Pa) at 25°C	S (mg L^{-1})	Log K_{OW}	Log K_{OC}	Log K_p	Log K_{ea}
Naphtalene	Nap	128	10.5	31.8	3.40	3	1.7	1.7
Acenaphtylene	Acy	152	-	3.93	3.61	-	-	-
Acenaphtene	Ace	154	0.356	3.7	3.77	3.7	-	-
Fluorene	Fl	166	0.09	1.98	3.96	3.9	-	-
Phenanthrene	Phe	178	0.018	1.2	4.57	4.2	2.73	2.8
Anthracene	Ant	178	$7.5.10^{-4}$	1.29	4.60	4.4	2.73	3.5
Fluoranthene	Flu	202	$1.2.10^{-3}$	0.26	5.1	4.9	3.7	3.7
Pyrene	Pyr	202	$8.86.10^{-4}$	0.13	5.32	-	3.7	-
Benz(a)anthracene	BaA	228	$7.3.10^{-6}$	0.011	5.31	-	4.19	-
Chrysene	Chry	228	$5.7.10^{-7}$	0.002	5.81	-	4.19	-
Benzo(b)fluoranthene	BbF	252	-	0.0012	6.57	5.8	5	3.3
Benzo(k)fluoranthene	BkF	252	6.10^{-7}	0.00076	6.84	5.9	4.88	-
Benzo(j)fluoranthene	BjF	252	-	0.0025	6.44	-	4.88	-
Benzo(e)pyrene	BeP	252	-	0.0063	6.44	-	4.88	-
Benzo(a)pyrene	BaP	252	$7.3.10^{-7}$	0.002	6.13	6	4.88	4.2
Perylene	Per	252	-	0.003	6.50	-	4.88	-
Dibenz(a,h)anthracene	DahA	278	$3.7.10^{-1}$	0.0005	6.7	6.1	5.65	5.6
Dibenz(a,c)anthracene	DacA	278	$1.3.10^{-9}$	0.023	7.1	-	5.65	-
Benzo(g,h,i)perylene	BP	276	-	0.00026	6.5	-	-	-
Indeno(1,2,3cd)pyrene	IP	276	-	0.062	6.6	6.8	5.57	-

Physical and chemical characteristics of PAHs generally vary with molecular weight. With increasing molecular weight, aqueous solubility decreases, and melting point, boiling point, and the log Kow (octanol/water partition coefficient) increases (Table 10), suggesting increased solubility in fats, a decrease in resistance to oxidation and reduction, and a decrease in vapor pressure. Accordingly, PAHs of different molecular weight vary substantially in their behavior and distribution in the environment and in their biological effects.

Source of Polycyclic Aromatic Hydrocarbons

Anthropogenic Sources of PAHs

Anthropogenic activities associated with significant production of PAHs include: coke production in the iron and steel industry; catalytic cracking in the petroleum industry; the manufacture of carbon black, coal tar pitch, and asphalt; heating and power generation; controlled refuse incineration; open burning; and emissions from internal combustion engines used in transportation.

Natural Source of PAHs

The PAHs that are present in the marine environment in relevant concentrations are divided into two groups depending on their origin, namely pyrogenic and petrogenic (Hylland et al. 2006). Pyrogenic PAHs are formed by incomplete combustion of organic material while the petrogenic PAHs are present in oil and some oil products (Feng et al. 2009). In general the pyrogenic PAHs are composed of larger ring systems, then the petrogenic PAHs. Sources of pyrogenic PAHs are forest fires incomplete combustion of fossil fuels and tobacco smoke. A range of PAHs are naturally present in crude oil and coal and these compounds are referred to as petrogenic PAHs.

Health Effect of PAHs

Several polycyclic aromatic hydrocarbons (PAHs) are among the most potent carcinogens known to exist, producing tumors in some organisms through single exposures to microgram quantities. The evidence implicating PAHs as an inducer of cancerous and precancerous lesions is becoming overwhelming, and this class of substances is probably a major contributor to the recent increase in cancer rates reported for industrialized nations In addition to the skin cancers noted initially, higher incidences of respiratory tract and upper gastrointestinal tract tumors were associated with occupational exposures to these carcinogens (Dipple et al. 1985). While some PAHs are potent mutagens and carcinogens, others are less active or suspected carcinogens. Some, especially those of biological origin, are probably not carcinogens. The US EPA has identified sixteen PAHs to seek priority; Fluorene, Fluoranthene and Pyrene. Depending on carcinogenicity proved or alleged, PAHs are classified differently. Table 11 presents the different classifications of the EU (Union European Commission), the International Agency for Research on Cancer (IARC) and the US EPA for PAH (Lafon et al. 2000). In the same table are also indicated their effects Harmful substances listed in the Register of Toxic Effects of Chemicals (RTEC).

Effect of PAH on Aquatic Organisms

PAHs are varying substantially in their toxicity to aquatic organisms (Table 5). In general, toxicity increases as molecular weight increases (although high molecular weight PAHs have low acute toxicity, perhaps due to their low solubility in water) and with increasing alkyl substitution on the aromatic ring. Toxicity is most pronounced among crustaceans and least among teleosts. An unusually high prevalence of oral, dermal, and hepatic neoplasms have been observed in bottom-dwelling fish from polluted sediments containing grossly-elevated PAH levels (Couch et al. 1985). While some PAHs are potent mutagens and carcinogens, others are less active or suspected carcinogens. Some, especially those of biological origin, are probably not carcinogens. Certain lower molecular weight, non carcinogenic PAHs, at environmentally realistic levels, were acutely toxic to aquatic organisms, or produced deleterious sublethal responses.

Table 11. Classification of 16 carcinogenic PAHs by different organizations (Topal et al. 2004)

Compounds	UE	CIRC IARC	US EPA	RTECS
Naphtalene			C	*T, M, R*
Acenaphtylene			*D*	*M*
Acenaphtene				*M*
Fluorene			*D*	*T, M*
Phenanthrene			*D*	*T, M, E*
Anthracene			*D*	*T, M, E*
Fluoranthene			*D*	*T, M, E*
Pyrene			*D*	*T, M, E*
Benzo(a)Annthracene		*2A*	*B2*	*T, M, C*
Chrysene	*Cat 2-R45*		*B2*	*T, M, E*
Benzo(b)Fluoranthene	*Cat 2-R45*	*2B*	*B2*	*T, M, C*
Benzo(k)Fluoranthene	*Cat 2-R45*	*2B*	*B2*	*T, M, E*
Benzo(a)Pyrene		*2A*	*B2*	*T, M, R, C*
Dibenzo(a,h)Anthracene	*Cat 2-R45*	*2A*	*B2*	*T, M, C*
Benzo(g,h,i)Pérylene			*D*	*T, M*
Indeno(1,2,3-cd)pyrene	*Cat 2-R45*	*2B*	*B2*	*T, M, C*

CIRC: 1: Carcinogenic to humans; 2A: Probably carcinogenic to humans 2B: Possibly carcinogenic to humans; 3:not classified

European Union: A: Carcinogenic to humans; B1 and B2: Probable carcinogen for humans; C: Possible carcinogen for humans; D: not classified; E: Probably not carcinogenic

US EPA: C: Carcinogen; Mutagenesis; A: Affects the reproductive system; T: Leads to the onset of tumors; E: Probable role in the appearance of tumors

European Substance Labeling: R45: May cause cancer RTECS: Cat 1: Carcinogenic to humans Cat 2: Substance classified as carcinogenic to humans Cat 3: Substance of human concern caused by carcinogenic effects Possible

Legislation and Convention for PAHs

Due to the toxic nature of PAHs, it is important to legislate on the content's maximum allowable to avoid environmental and human risk. Regarding the waters, The World Health Organization (WHO) defines limits for drinking water to 5 µg L^{-1} for Fluoranthene and 0.7 µg L^{-1} for Benzo (a) Pyrene. At present, there are no regulations in France, as in Tunisia, Levels of PAH in sediments. In Tunisia, the only available data on the oil pollution is provided by the National Agency of Environmental Protection (ANPE 1990).

Example of study 1: Distribution of polycyclic aromatic hydrocarbons in sediment cores from the Sicily Channel and the Gulf of Tunis (south-western Mediterranean Sea) (Mzoughi et al. 2013)

In this study total PAH refers to the sum of 24 individual compounds analyzed. Total PAH concentration in the sediment cores ranged from 29.7 to 405.7 ng g^{-1} in the Sicily Channel and from 32.3 to 709.2 ng g^{-1} in the Gulf of Tunis (Table 12). Figure 4 presented study area and core sample locations St 1 and St 2.The highest concentrations were found at the depth of 2 cm in the two sediment cores. The second peak was found at the 10 cm depth, with concentrations of 356.8 ng g^{-1} and 160.6 ng g^{-1}, respectively in sediment cores of the Sicily Channel and the Gulf of Tunis. At the sediment surface, the total PAH concentration was 342.8 ng g^{-1} in the Sicily Channel and 100.2 ng g^{-1} in the Gulf of Tunis. According to Baumard et al. (Baumard et al. 1998), PAH levels can be characterized as low, moderate, high and very high when PAH concentrations are 0–100, 100–1000, 1000– 5000 and >5000 ng g^{-1} , respectively. On the basis of classification adopted by Baumard et al. (Baumard et al. 1998), the sediments from the Gulf of Tunis and the Sicily Channel can be considered low to moderate polluted with PAH, although they would be influenced by the substantial traffic of deep-sea fishing vessels and other ships traveling along the navigational route.

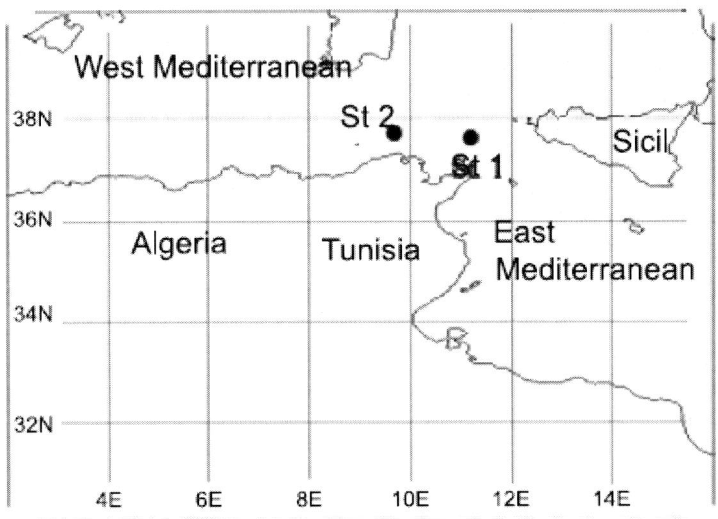

Figure 4. Study area and core sample locations (St 1 and St 2).

Example of study 2: Distribution and partitioning of aliphatic hydrocarbons and polycyclic aromatic hydrocarbons in sediment from the harbors of the West coast of the Gulf of Tunis (Tunisia) (Mzoughi et al. 2011)

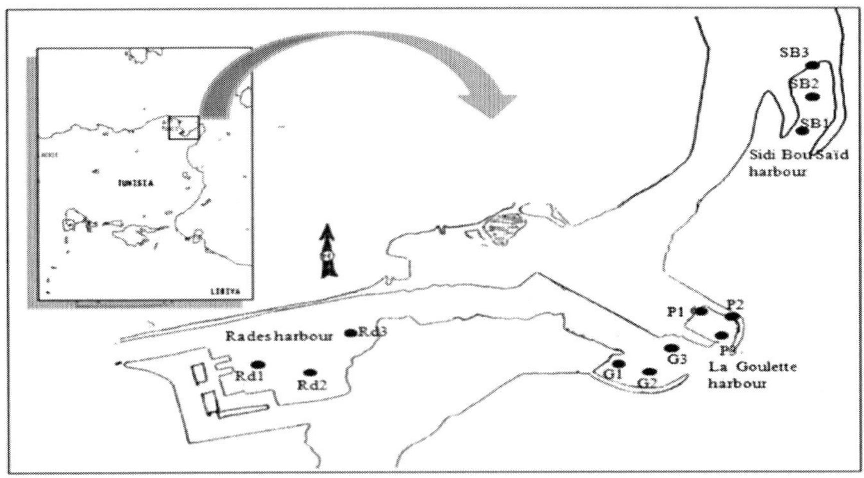

Figure 5. Location map of studied samples.

Table 12. Concentration of total aliphatic hydrocarbons (AH), total polycyclic aromatic hydrocarbons (PAH) (ng g^{-1}) and different ratios characterizing the origin of hydrocarbons

Depth (cm)	Total AH		Pr/ph		ANT/ANT+PHN		Total PAH		FLR/FLR+PYR		B(a)ANT/B(a)ANT+Chr	
	(a)	(b)	(a)	(b)	(a)	(b)	(a)	(b)	(a)	(b)	(a)	(b)
1	563.3	2245.7	1.87	2.46	0.98	0.97	342.9	106.4	0.26	1	0.6	1
2	1838.2	2767.1	0.28	3.43	0.99	0.99	290	618.1	0.57	0.33	0.53	0
3	922.9	1172.1	1.58	3.22	0.99	0.95	316.3	246.7	0.75	0.3	0	0.85
5	232.6	988.6	0	2.27	0.99	0.99	239.8	82.5	0.35	1	0.64	1
7	340.5	1601.1	0	1.71	0.98	0.98	188.6	56.4	0.03	0.007	0.78	0
10	1036.4	1889.9	1.41	1.02	0.99	1	364.3	185.9	0.02	0.006	0.47	1
13	240.7	1034.5	0	0.82	0.99	0	209.03	14.7	0	1	0.43	0
17	240.7	734.9	1.14	0.46	0.99	1	222.45	87.3	0	0	0.63	0
20	285.1	1806.5	0.83	1.78	1	0.98	26.89	47.8	0	0	0.4	0

Pr/ph: pristane to phytane; ANT/ANT+PHN: anthracene to anthracene + phenanthrene; FLR/FLR+PYR: fluoranthene to fluoranthene + pyrene; B(a)ANT/B(a)ANT+Chr: benz(a)anthracene to benz(a)anthracene + chrysene (a): Sicily Channel; (b): Gulf of Tunis

Table 13. Concentration of aliphatic (AH), polycyclic aromatic (PAH) in sediment (ng g^{-1}) and different ratios characterizing the origin of hydrocarbons

	Season	Rd1	Rd2	Rd3	G1	G2	G3	P1	P2	P3	SB1	SB2	SB3
Pr/Ph	Summer	0.6	0.9	1.8	0.9	0.8	0.7	0.8	0.6	0.9	0	0.7	0.7
	Winter	0.6	0.6	3.1	1	0.7	0.8	0.5	0.4	0.6	0.8	0.3	0.8
	Average	0.6	0.8	2.5	1	0.8	0.8	0.7	0.5	0.8	0.4	0.5	0.8
PAH	Summer	4944	4438.8	5816.2	5124	3572.7	5373	4513.7	9948.1	5917.3	2091.2	2329.5	989
	Winter	6968.3	4152	3756.2	2816.1	5728.4	5656.2	9511.1	7203.5	5759.4	2706.5	3596.2	1802
	Average	5956.1	4295.4	4786.2	3970	4650.5	5514.6	7012.4	8575.8	5838.3	2398.9	2962.8	1395.5
PPAH	Summer	1170.4	1252	648.3	1093.9	1423	564.1	4358	445.7	717	1828.4	1087.2	947.6
	Winter	1411.2	451.9	547.6	3330.2	623.5	497.1	1821.2	7026.4	1002.2	504.2	363.3	512.6
	Average	1290.8	852	598	2212	1023.3	530.6	3089.6	3736	859.6	1166.3	725.3	730.1
Phe/An	Summer	0.1	0	3	0.2	1.9	0	1.7	0	2.3	0	0.3	4
	Winter	0.6	0.9	0.9	0.9	1.2	1.1	0.4	9.6	0.7	0.8	0.8	1.2
	Average	0.4	0.4	1.9	0.5	1.6	0.6	1.1	9.6	1.5	0.4	0.6	2.6
Fl/Py	Summer	2.3	0.2	0	4.9	1.2	0.1	0.2	0.7	1	1	0.1	0.1
	Winter	0.7	0.2	0.2	1	1.7	0.8	0.1	2.6	0.1	1.5	0.2	0.3
	Average	1.5	0.2	0.1	2.9	1.4	0.5	0.2	1.6	0.5	1.3	0.2	0.2
BaP/BeP	Summer	0.3	0.8	2	0.9	0	0.4	1.2	0	0.8	0.3	1.4	1.2
	Winter	2.7	2	7	0.7	3.2	2.3	2.2	3.8	1.5	1.2	2.2	1.5
BaA/Chr	Summer	0	0	0.1	0.6	1	0.2	0.4	0.7	0.3	0.4	0.1	0
	Winter	2.5	1	2.7	1.1	5.3	0.4	2	0.8	0.7	0.5	0.7	0
Total TEQ	Summer	583.8	133.3	35.2	201.6	232.2	8.2	65.6	9.4	14.2	126.8	83.4	25.8
	Winter	46.3	25.8	74.5	666.4	49.3	58	104.2	55.5	80.8	20.7	38.3	25.8
	Average	315.1	79.6	54.9	434	140.8	33.1	84.9	32.4	47.5	73.7	60.9	141

Pristane to phytane: Pr/Ph, phenanthrene to anthracene: Phe/An, fluoranthene to pyrene: Fl/Py, benz(a)pyrene to benz(e)pyrene: BaP/BeP, benz(a)anthracene to chrysene: BaA/Chr and Total toxic benzo(a)pyrene equivalent: Total TEQ. Rades (Rd), La Goulette fishing (P) and passenger (G), and Sidi Bou Said (SB)

Concentrations of AH in sediment varied from 1802 ng g^{-1} dw (dry weight) to 9511.1 ng g^{-1} dw in winter and from 989 ng g^{-1} dw to 9948.1 ng g^{-1} dw in summer (Table 13). Figure 5 presents the location map of studied samples. The highest concentration of PAH was recorded for station P2 of La Goulette fishing harbor and the lowest level of PAH was recorded for station SB3 of Sidi Bou Said harbor in summer and in winter. The percentage of aliphatic fraction (F1) varied from 46% to 91% of total hydrocarbons (average 80%) in winter and from 51% to 96% of total hydrocarbons (average 75%) in summer (significantly different $p < 0.05$). Concentrations of total PAH in sediment ranged from 363.3 ng g^{-1} dw to 7026.4 ng g^{-1} dw in winter and from 445.7 ng g^{-1} dw to 4358 ng g^{-1} dw in summer. PAHs represent 8% to 54% of total hydrocarbons (average 20%) in winter and 4% to 49% of total hydrocarbons (average 25%) in summer (Table 13). Percentages of PAH found in summer and in winter were statistically significantly different ($p < 0.05$).

The highest concentrations of PAH in summer and in winter were registered in sediment from La Goulette (P1) harbor which could be attributed to several reasons. Preferential retaining of PAH by high TOC in the sediments at station P1 of La Goulette fishing harbor relative to Sidi Bou Said harbor may have resulted in higher PAH concentration in sediment samples probably related to diminished inputs of PAH which resulted in lower PAH concentrations for stations of Sidi Bou Said harbor than stations of La Goulette fishing harbor. The composition of PAH was variable in sediment samples. The PAH with 2,3-rings ranged from 8.1 to 45.4% (average 32.5%) of the total PAH in summer and ranged from 10.4 to 72.5% (average 42.2%) of the total PAH in winter. The percentage of PAH with 4 rings varied between 3 and 80.3% (average 35.9%) of the total PAH in summer and between 6.2 and 27.7% (average 17%) of the total PAH in winter. The same result was obtained for PAH with 5.6 rings with an average of 31.5% in summer and 40.8% in winter (Figure 5). Several PAHs, and especially their metabolic products, are known to be carcinogenic. The total concentration of potentially carcinogenic PAH (CPAH) (sum of benzo(a)anthracene, benzo(b)fluoranthene, benzo(k) fluoranthene, benzo(a)pyrene, indeno(1,2,3-cd)pyrene and dibenzo(a,h) anthracene) varied from 47.6 to 1446.3 ng g^{-1}, accounted for 3.3–43.4% of total sediment PAH in winter and from 36 to 164.6 ng g^{-1}, accounted for 3–31.5% total sediment PAH in summer. Total TEQ calculated for all samples investigated varied from 20.7 to 666.4 ng g^{-1} TEQ in winter and from 8.2 to 583.8 ng g^{-1} TEQ in summer (Table 13). The maximum value of Total TEQ was found at station G1 of La Goulette passenger harbor in winter and at station Rd1 of Rades harbor (industrial area) in summer indicating the presence of local source of contamination.

Heavy Metals (HM)

Definition

Heavy metals, as defined by Neiboer and Richardson are normal constituents of the marine environment. At least 10 heavy metals are known to be essential to marine organisms: iron (Fe), copper (Cu), zinc (Zn), cobalt (Co), manganese (Mn), chromium (Cr), molybdenum (Mo), vanadium (V), selenium (Se) and Nickel (Ni). These metals always function in combination with organic molecules, usually proteins. Metals occur normally at low concentrations yet are capable of exerting considerable biological effects, even at such levels (Kirpichtchikova et al. 2006). All metals are toxic at levels higher than threshold limit. Silver (Ag), mercury (Hg), copper (Cu), cadmium (Cd) and lead (Pb) are particularly toxic. Metal pollution of the sea is less visible and direct than other types of marine pollution, but its effects on marine ecosystems and humans are intensive and very extensive. Table 14 presents physical and chemical properties of heavy metals.

Sources and Emissions of Heavy Metals

Natural Sources of Heavy Metals

The natural sources of atmospheric heavy metal contamination include mainly the dried water droplets from the oceans, dust particles from volcanoes, erosion of soil, weathering of rocks and forest fires. Biodegradation of dead animals and plants also contribute significantly to background levels of metals in waters. Heavy metal concentration in waters can be attributed to both natural and anthropogenic sources. The most significant source of metals is the weathering of rocks and volcanic activities from which the released metals find their way into the water bodies. A large quantity of metals also suspends into the atmosphere from where they can reach the waters through dry deposition and with the rainfall (Agarwa et al. 2009).

Table 14. Physical and chemical properties of heavy metals

Metal	group	period	atomic number	atomic mass	Density g cm^{-3}	melting point°C	boiling point°C
Lead	IV	6	82	207.2	11.4	327.4	1725
Chromium	VIB	4	24	52	7.19	1875	2665
Arsenic	VA	4	33	75	5.72	817	613
Zinc	IIB	4	30	65.4	7.14	419.5	906
Cadmium	XII	5	48	112.4	8.65	320.9	765
Copper	IB	4	29	63.5	8.96	1083	2595
Mercury	XII	6	80	200.6	13.6	13.6	357
Nickel	X	4	28	58.69	8.908	1453	2732

Anthropogenic Sources of Heavy Metals

Heavy metals are often problematic environmental pollutants, with well-known toxic effects on living systems. Nevertheless, because of certain useful physical and chemical properties, some heavy metals, including mercury, leads, and cadmium, are intentionally added to certain consumer and industrial products such as batteries, switches, circuit boards, and certain pigments. Usage patterns of heavy metals in products have varied over the years. Heavy metals, to a large extent, are dispersed in the environment through industrial effluents, organic wastes, refuse burning, transport and power generation. They can be carried to places far away from the sources by wind, depending upon whether they are in gaseous form or particulates. Metallic pollutants are ultimately wasted out of the air, rain into the land or the surface waters. Metal containing industrial effluents constitute a major source of metallic pollution of the hydrosphere. Another means of dispersal in the movement of drainage water from catchment areas with have been contaminated by waste from mining and smelting units.

Application of Heavy Metals

Lead was used in industrial production includes soldiers, bearings, cable covers, ammunition, plumbing, pigments, and caulking (Manahan et al. 2003). Zinc is added during industrial activities, such as mining, coal, and waste combustion and steel processing. Many foodstuffs contain certain concentrations of zinc. Drinking water also contains certain amounts of zinc, which may be higher when it is stored in metal tanks. The most significant use of the cadmium is in nickel/cadmium batteries, as rechargeable or secondary power sources exhibiting high output, long life, low maintenance, and high tolerance to physical and electrical stress. Cadmium coatings provide good corrosion resistance coating to vessels and other vehicles, particularly in high-stress environments such as marine and aerospace. Other uses of cadmium are as pigments, stabilizers for polyvinyl chloride (PVC), in alloys and electronic compounds. Cadmium is also present as an impurity in several products, including phosphate fertilizers, detergents and refined petroleum products (Campbell et al. 2006). Cadmium is produced as an inevitable byproduct of zinc and occasionally lead refining (Weggler et al. 2004). The most common application of nickel is an ingredient of steel and other metal products (Khodadoust et al. 2004). Mercury was once widely used in pharmaceutical products, agricultural chemicals, dry cell batteries, and paints. Many of these uses have been phased out, although others continue, e.g., Chlor-alkali production, switches and electrical apparatus, fluorescent light bulbs, and dental amalgam (Sardar et al. 2013).

Health Effect of Heavy Metals

The biotoxic effects of heavy metals refer to the harmful effects of heavy metals in the body when consumed above the bio-recommended limits. Inhalation and ingestion are the two routes of exposure, and the effects of both are the same. Lead accumulates in the body organs (i.e., brain), which may lead to poisoning (plumbism) or even death. The gastrointestinal tract, kidneys, and central nervous system are also affected by the presence of lead. Children exposed to lead are at risk for impaired development, shortened attention span, hyperactivity, and mental deterioration, with children under the age of six being at a more substantial risk. Adults usually experience decreased reaction time, loss of memory, nausea, insomnia, anorexia, and weakness of the joints when exposed to lead. Lead is not an essential element. It is well known to be toxic and its effects have been more extensively reviewed than the effects of other trace metals. Lead can cause serious injury to the brain, nervous system, red blood cells, and kidneys (Baldwin et al. 1999). Exposure to lead can result in a wide range of biological effects depending on the level and duration of exposure. Various effects occur over a broad range of doses, with the developing young and infants being more sensitive than adults. Chromium is associated with allergic dermatitis in humans (Scragg et al. 2006). Arsenic is associated with skin damage, increased risk of cancer, and problems with the circulatory system (Scragg et al. 2006). Nickel is an element that occurs in the environment only at very low levels and is essential in small doses, but it can be dangerous when the maximum tolerable amounts are exceeded. Zinc is a trace element that is essential for human health. Zinc shortages can cause birth defects. Cadmium is very persistent but has few toxicological properties and, once absorbed by an organism, remains resident for many years. Cadmium in the body is known to affect several enzymes. It is believed that the renal damage that results in proteinuria is the result of Cd adversely affecting enzymes responsible for absorption of proteins in kidney tubules. Cadmium also reduces the activity of delta-aminolevulinic acid synthetase, arylsulfatase, alcohol dehydrogenase, and lipoamide dehydrogenase, whereas it enhances the activity of deltaaminolevulinic acid dehydratase, pyruvate dehydrogenase, and pyruvate decarboxylase (Manahan et al. 2003). The major threat to human health is chronic accumulation in the kidneys leading to kidney dysfunction. Food intake and tobacco smoking are the main routes by which cadmium enters the body (Manahan et al. 2003). Copper is an essential micronutrient required in the growth of both plants and animals. In humans, it helps in the production of blood hemoglobin. Copper is indeed essential, but in high doses it can cause anemia, liver and kidney damage, and stomach and intestinal irritation (Bjuhr et al. 2007). Mercury is associated with kidney damage. This can cause various kinds of cancer on different sites in the body, mainly of those that live near refineries.

Effects of Heavy Metals on Aquatic Organisms

Inhibition of growth is one of the most distinct symptoms of toxic action of metals on fish larvae. Therefore, the fish body length and mass are indicators of environmental conditions. Results of a research showed that the highest concentration of heavy metals is in kidney and liver of ten different fish species. Contaminated sediments can threaten creatures in the benthic environment, exposing worms, crustaceans and insects to hazardous concentrations of toxic chemicals. Some kinds of toxic sediments killed the benthic organisms, and reduced the food availability for larger animals such as fish. (GWRTAC. 1997).

Legislation and Convention for Heavy Metals

The Executive Body adopted the Protocol on Heavy Metals in Aarhus (Denmark) on 24 June 1998. It targets three particularly harmful metals: cadmium, lead and mercury. In 2012, Parties to the Protocol on Heavy Metals adopted a decision to amend the Protocol to include more stringent controls of heavy metals emissions and to introduce flexibilities to facilitate accession of new Parties, notably countries in Eastern Europe, Southeastern Europe, the Caucasus and Central Asia. In 2013, the Minamata Convention on Mercury was adopted, a treaty negotiated under the auspices of the United Nations Environment Programme (UNEP). Building on the 1998 Protocol on Heavy Metals, the Minamata Convention raised the profile of mercury to the global level.

Example of study 1: Distribution of trace metals in sediment cores from the Sicily Channel and the Gulf of Tunis (south-western Mediterranean Sea) (Mzoughi et al. 2011)

Similar profiles for total metal in the two cores were observed for Cd and Hg (Figure 6). Figure 4 presents study area and core sample locations St 1 and St 2. Metal concentrations were lowest in the deepest part of the core, increased rapidly up to approximately 7 cm depth for Cd and up to 5 and 10 cm for mercury, and decreased slightly towards the surface. Profiles of Pb, Fe, Cu, Zn and Mn were quite similar and varied little with depth in sediment cores, revealing a more or less uniform distribution throughout the length, suggesting similar sediment texture and character (Ram et al. 2003). Cadmium and Hg displayed a small increase at subsurface depths, with a peak around 7 cm depth for Cd and 2, 5 and 10 cm depths for Hg, whereas Pb concentrations were considerably higher and remained constant or decreased gradually with increasing core depth. Surface sediment distributions can change through time and these variations are influenced by hydrodynamics as well as biological and chemical processes. The ranking of abundance based on absolute concentrations of the metals from the Sicily Channel and the Gulf of Tunis is Mn > Zn > Fe > Cu > Pb > Cd > Hg. In this hierarchy, Mn represents four times the Zn concentration in the Sicily Channel and eight times the Zn concentration in the Gulf of Tunis, whereas the Cd levels are approximately 200 times lower than Pb concentrations. The values for Mn, Cu and Pb in the sediment core of the Gulf of Tunis are two times greater than those from the Sicily Channel whereas, for the rest of the metals analyzed, the values are approximately the same. Chemical partitioning shows that the enrichment in the surface and near-surface sediments are related to the relatively high proportion of the total metal concentrations that occur in the exchangeable fraction, and they generally decrease with depth. Copper, Pb and Zn likely derive from those metals held in ion exchange positions, certain carbonates, and from easily soluble amorphous compounds of Mn and perhaps those of Fe. Sirinawin and Sampongchaiyakul (Sirinawin et al. 2005) showed that diagenetic processes involving Mn, as well as the changes in the oxidizing/reducing boundaries within core depth, appear to be the most important factors in controlling the behavior of the metals in a sediment core.

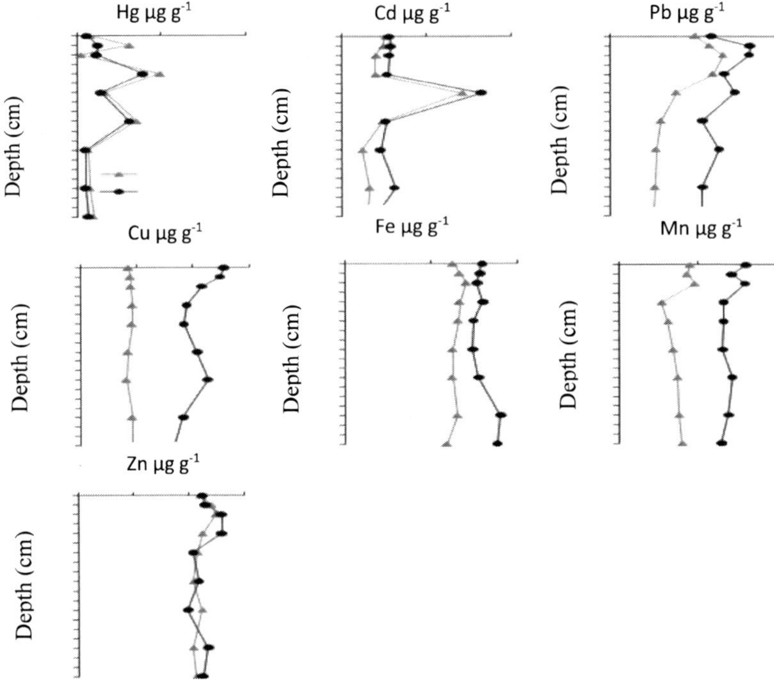

Figure 6. Vertical depth profiles of Cd, Hg, Pb, Fe, Cu, Zn and Mn (µg g^{-1} dry weight) in two sediment cores collected from the Sicily Channel and the Gulf of Tunis.

Example of study 2: Heavy Metals and PAH Assessment Based on Mussel Caging in the North Coast of Tunisia (Mediterranean Sea) (Mzoughi et al. 2012)

The ranges of trace metals concentrations expressed in µg/gdw are: Hg (0.1-0.2), Pb (0.4-0.7), Cd (0.9-2.9), Cu (2.9-3.9), Fe (117-248) and Zn (250-426). Higher concentrations were observed at Rades, La Galite and Tabarka which can be attributed to the industrial activities implanted in the coast of Algeria and in the city of Rades (Table 15). Concentrations of Cd, Pb, Fe and Zn are generally different from initial concentrations (before caging), depending on the adaptability of transplanted mussels to be used as bio-indicator of contaminants. Yet bio-monitoring using mussels give information on compound bioavailability, which depends on their ability to accumulate contaminants in its tissue. Concentrations of all measured metals were higher than the initial mussel samples of Languedoc-Roussillon at t = 0. No difference was obtained for Cu in all stations at t = 0 and at t = 3 months. Some concentrations of bioaccumulation heavy metals (Cd, Pb, Cu, Fe and Zn) were higher at Tabarka, La Galite and Rades stations (figure 7). Contaminant concentrations obtained during this study were compared with those obtained in the context of Network of harmful contaminants (RECNO) performed at various sites along the Tunisian Mediterranean coast between 1996 and 2010.

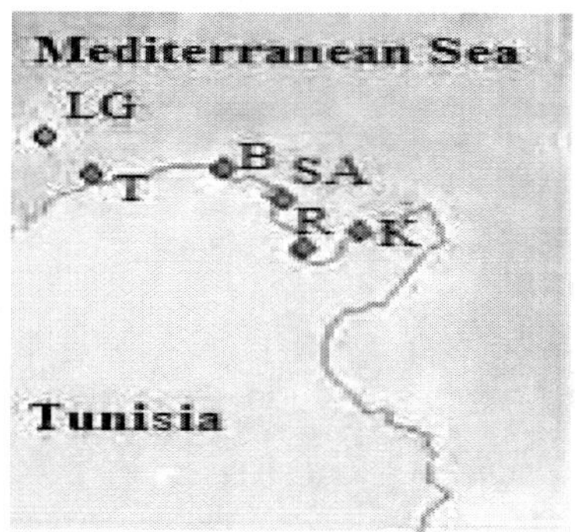

Figure 7. Map of the stations of caged mussels in the North coast of Tunisia (LG: Languedoc-Roussillon, T: Tabarka, B: Bizerte, SA: Sidi Ali, R: Rades and K: Korbous).

Table 15. Concentration of Hg, Cd, Pb, Cu, Fe and Zn (µg g^{-1} dw) in caged mussels from different area of the North coast of Tunisia

Area	Hg	Cd	Pb	Cu	Fe	Zn
Languedoc-Roussillon	0.12	0.87	0.36	3.89	142	256
Tabarka	0.2	2.94	0.74	3.61	247	426
La Galite	0.15	2.18	0.54	3.61	117	314
Bizerte Lagoon	0.15	1.79	0.65	3.32	184	299
Sidi Ali	0.22	1.74	0.45	3.62	145	251
Rades	0.13	1.68	0.63	2.92	211	316
Korbous	0.21	1.79	0.54	3.54	248	274

They highlighted approximately the same levels. Heavy metals are in solute form and homogeneous levels in the water column. These metals seem to have little impact on variations in levels between marine stations. Under these conditions, levels of metals measured in natural populations sampled on the coast are nearly identical to those obtained from transplantation of caged mussels. Heavy metal concentrations can be either natural due to geological substrates or man induced, due to direct or indirect input related to human activities. Consequently, the natural occurrence of metals complicates assessments of potentially contaminated sites because measurable metal bioavailability does not automatically infer contamination and some site specific high concentrations may represent entirely natural conditions.

REFERENCES

Adeyemi, D. Anyakora, C. Ukpo, G. Adedayo, A. (2011). Evaluation of the levels of organochlorine pesticide residues in water samples of Lagos Lagoon using solid phase extraction method, *J. Environ. Chem. Ecotoxicol,* 3, 160-166.

Agarwal, S. K. (2009), *Heavy Metal Pollution*, 1st edition 161-165.

ANPE (1990). Étude préliminaire de l'écologie du lac de Bizerte. Rapport, Agence Nationale de Protection de l'Environnement, Tunisie, 100. [Preliminary study of the ecology of Lake Bizerte. Report, National Agency Of Environment Protection, Tunisia, 100].

Association for Environmental Protection in Kerkennah (APEK) (2005) Tunisia Country Situation Report.

Baldwin, D. R., Marshall, W. J. (1999). Heavy Metal Poisoning and its Laboratory Investigation. *Ann. Clin. Biochem.* 36, 267-300.

Ballschmiter, K., Zell, M. (1980). Analysis of polychlorinated biphenyls (PCB) by glass capillary gas chromatography. *Fresenius' Zeitschrift Für Analytische Chemie*, *302*, 20-31.

Baumard, P., Budzinski, H., Garrigues, P. (1998). PAHs in Arcachon Bay, France: Origin and biomonitoring with caged organisms. *Mar. Pollut. Bull, 36*, 577-586.

Bergman, Å., Rydén, A., Law, R. J., de Boer, J., Covaci, A., Alaee, M., van der Veen, I. (2012). A novel abbreviation standard for organobromine, organochlorine and organophosphorus flame retardants and some characteristics of the chemicals. *Environ. Int,* 49, 57-82.

Bjuhr, J. (2007) Trace Metals in Soils Irrigated with Waste Water in a Periurban Area Downstream Hanoi City, Vietnam.

Breivik, K. Gioia, R. Chakraborty, P. Zhang, G. Jones, K.C. (2011). Are reductions in industrial organic contaminants emissions in rich countries achieved partly by export of toxic wastes? *Environ. Sci. Technol.,* 45, 9154–9160.

Campbell, P. G. C. (2006). "Cadmium-A priority pollutant," *Environmental Chemistry*, 3, 387–388,

Couch, J. A., Harshbarger, J. C. (1985). Effects of carcinogenic agents on aquatic animals: An environmental and experimental overview. *J. Environ. Sci. Health. C Environ. Carcinog. Ecotoxicol. Rev.*, 3, 63-105.

Council Directive 98/83/EC (1998) on the quality of water intended for human consumption, of. *J. Eur. Commun.*

Dipple, A. (1985). Polycyclic Aromatic Hydrocarbon Carcinogenesis, *J. Am. Chem. Soc.* 283, 1-17.

Feng, X., Pisula, W., Müllen, K. (2009). *Large polycyclic aromatic hydrocarbons: Synthesis and discotic organization.*

Garabrant, D. H., Held, J., Langholz, B., Peters, J. M., Mack, T. M. (1992). DDT and related compounds and risk of pancreatic cancer. *J. Natl. Cancer Inst*, 84, 764-771.

Gao, J. Liu, L. Liu, X, Lu, J. Zhou, H. Huang, S. Wang, Z. Spear P.A. (2008). Occurrence and distribution of organochlorine pesticides – lindane, p,p′-DDT, and heptachlor epoxide – in surface water of China, *Environ. Int,* 34, 1097–1103.

Gioia, R., Akindele, A. J., Adebusoye, S. A., Asante, K. A., Tanabe, S., Buekens, A., Sasco, A. J. (2014). Polychlorinated biphenyls (PCBs) in Africa: a review of environmental levels. *Environ. Sci. Pollut. Res. Int.*, 21, 6278-6289.

GWRTAC (1997). "Remediation of metals-contaminated soils and groundwater," Tech. Rep. TE-97-01.

Haib, J., Hofer, I., Renaud, J. M. (2003). Analysis of multiple pesticide residues in tobacco using pressurized liquid extraction, automated solid-phase extraction clean-up and gas chromatography-tandem mass spectrometry. *J. Chromatogra. A*, *1020*, 173-187.

Hylland, K. (2006). Polycyclic aromatic hydrocarbon (PAH) ecotoxicology in marine ecosystems. *J. Toxicol. Environ. Health.* A. 69, 109-123.

Jiang, X., Xia, Z., Deng, L., Wei, W., Chen, J., Xu, J., Li, H. (2012). Evaluation of Accuracy for the Measurement of Octanol–Water Partition Coefficient by MEEKC. *Chromatographia*, *75*, 347-352.

Jurewicz, J., Hanke, W. (2008). Prenatal and Childhood Exposure to Pesticides and Neurobehavioral Development: Review of Epidemiological Studies. *Int. J. Occup. Environ. Med. 21*, 121–132.

Khodadoust, A. P., Reddy, K. R., Maturi, K. (2004). Removal of Nickel and Phenanthrene from Kaolin Soil Using Different Extractants. *Environ. Eng. Sci., 21*, 691-704.

Kirpichtchikova, T. A. Manceau, A. Spadini, L. Panfili, F. Marcus, M.A. Jacquet, T. (2006). "Speciation and solubility of heavy metals in contaminated soil using X-ray microfluorescence, EXAFS spectroscopy, chemical extraction, and thermodynamic modeling," Geochim. Cosmochim. Acta. 70, 2163–2190,

Kuranchie-Mensah, H., Atiemo, S. M., Palm, L. M. N. D., Blankson-Arthur, S., Tutu, A. O., Fosu, P. (2012). Determination of organochlorine pesticide residue in sediment and water from the Densu river basin, Ghana. *Chemosphere*, *86*, 286-292.

Lafon, D., Pichard, A., Bisson, M., (2000). Évaluation du danger toxicologique du fioul rejeté sur les côtes. INERIS, Dossier ERIKA, Rapport 3. Institut National de l'Environnement Industriel et des Risques, Verneuil-en-Halatte, France. [Assessment of the toxicological hazard of fuel oil discharged to the coasts. INERIS, File ERIKA Report 3. National Institute for Industrial Environment and Risks, Verneuil-en-Halatte, France].

Lau, M. H. Y Leung, K. M. Y. Wong, S. W. Y. Wang, H. Yan, Z. G. (2012). Environmental policy, legislation, and management of persistent organic pollutants (POPs) in China, *Environ. Pollut,* 165, 182–192.

Liu, W. X., He, W., Qin, N., Kong, X. Z., He, Q. S., Ouyang, H. L., Xu, F. L. (2013). The residues, distribution, and partition of organochlorine pesticides in the water, suspended solids, and sediments from a large Chinese lake (Lake Chaohu) during the high water level period. *Environ. Sci. Pollut. Res. Int. 20*, 2033-2045.

Manahan, S. E. (2003). Toxicological Chemistry and Biochemistry, CRC Press, Limited Liability Company (LLC), 3rd edition.

Mastral, A. M., Callén, M. S. (2000). A Review on Polycyclic Aromatic Hydrocarbon (PAH) Emissions from Energy Generation. *Environ. Sci. Technol.*, *34*, 3051-3057.

Meador, J. P. (1996). Environmental Contaminants in Wildlife: Interpreting Tissue Concentrations, 117-153.

Mzoughi, N. Chouba, L. (2011). Distribution of trace metals, aliphatic hydrocarbons and polycyclic aromatic hydrocarbons in sediment cores from the Sicily Channel and the Gulf of Tunis (south-western Mediterranean Sea). *Environ. Technol.,* 32, 43–54.

Mzoughi, N. Chouba, L. (2011). Distribution and partitioning of aliphatic hydrocarbons and polycyclic aromatic hydrocarbons between water, suspended particulate matter, and sediment in harbours of the West coastal of the Gulf of Tunis (Tunisia). *J. Environ. Monit.,* 13, 689–698.

Mzoughi, N. Chouba, L. (2012). Heavy Metals and PAH Assessment Based on Mussel Caging in the North Coast of Tunisia (Mediterranean Sea). *Int. J. Environ. Res.,* 6, 109-118,

Necibi, M., Mzoughi, N., Yahia, M. N. D., Pringault, O. (2015). Distributions of organochlorine pesticides and polychlorinated biphenyl in surface water from Bizerte Lagoon, Tunisia. *Desalin. Water. Treat.*, *56*, 1-9.

Nunez, O., Moyano, E., Galceran, M. T. (2005). LC–MS/MS analysis of organic toxics in food. *Trends Analyt. Chem. 24*, 683-703.

Palanza, P., Morellini, F., Parmigiani, S., vom Saal, F. S. (1999). Prenatal exposure to endocrine disrupting chemicals: effects on behavioral development. *Neurosci. Biobehav. Rev., 23*, 1011-1027.

Ram, A., Rokade, M. A., Borole, D. V., Zingde, M. D. (2003). Mercury in sediments of Ulhas estuary. *Mar. Pollut. Bull*, *46*, 846-857.

Robertson, L. W., Hansen, L. G. (2015). PCBs: Recent Advances in Environmental Toxicology and Health Effects. University Press of Kentucky.

Ritter, L. Solomon, K. R. (1995) les pollutants organiques persistents les polluants organiques persistants: Rapport d'évaluation DDT-aldrine-dieldrine-endrine-chlordane Heptachlore-hexachlorobenzène Mirex-toxaphène Biphényles polychlorés Dioxines et furanes 7-54.

Sardar, K. Ali, S. Hameed, S. Afzal, S. Fatima, S. Shakoor, M. B, (2013). Heavy metals contamination and what are the impacts on living organisms, 2, 172-179.

Scragg, A. (2006). Environmental Biotechnology, Oxford University Press, Oxford, UK, 2nd edition,

Sirinawin W. Sampongchaiyakul, P. (2005). Nondetrital and total metal distribution in core sediments from the U-Tapao Canal, Songkhla, Thailand, *Mar. Chem.*, 94, 5–16.

Shen, L. Wania, F. Compilation, (2005). Evaluation, and Selection of Physical-Chemical Property Data for Organochlorine Pesticides, *J. Chem. Eng. Data* 50, 742-768.

Topal, M. H., Wang, J., Levendis, Y. A., Carlson, J. B., Jordan, J. (2004). PAH and other emissions from burning of J P-8 and diesel fuels in diffusion flames. Fuel 83, 2357–2368.

Van Ginneken, V., Palstra, A., Leonards, P., Nieveen, M., van den Berg, H., Flik, G., Murk, A. (2009). PCBs and the energy cost of migration in the European eel (Anguilla anguilla L.). *Aquat. Toxicol., 92*, 213-220.

Weggler, K., McLaughlin, M. J., Graham, R. D. (2004). Effect of chloride in soil solution on the plant availability of biosolid-borne cadmium. *J. Environ Qual., 33*, 496-504.

Wurl, O., Obbard, J. P. (2005). Chlorinated pesticides and PCBs in the sea-surface microlayer and seawater samples of Singapore. *Mar. Pollut. Bull, 50*, 1233-1243.

Yang, Z. Shen, Z. Gao, F. Tang, Z. (2009) Occurrence and possible sources of polychlorinated biphenyls in surface sediments from the Wuhan reach of the Yangtze River, China, *Chemosphere,* 74, 1522–1530.

Yum, S., Woo, S.; Kagami, Y., Park, H. S., Ryu, J. C., (2010). "Changes in gene expression profile of medaka with acute toxicity of Arochlor 1260, a polychlorinated biphenyl mixture." Comparative Biochemistry and Physiology. *Comp. Biochem. Physiol. C: Pharmacol. Toxicol.* 151, 51–56.

Zitko, V, (2002). Chlorinated Pesticides: Aldrin, DDT, Endrin, Dieldrin, Mirex. *The Handbook of Environmental Chemistry*, 30, 47-90.

In: Micropollutants
Editor: Tabitha N. Holloway

ISBN: 978-1-53612-067-7

Chapter 2

EFFECT OF MANGANESE AND FERRIC IONS ON THE DEGRADATION OF DI-2-ETHYLHEXYL PHTHALATE (DEHP) BY *ACINETOBACTER* SP. SN13

Renata Alves de Toledo, Jiaming Xu, U. Hin Chao and Hojae Shim*

Department of Civil and Environmental Engineering,
Faculty of Science and Technology,
University of Macau, Macau, SAR, China

ABSTRACT

Phthalate esters (PAEs) are artificially synthetized organic compounds extensively used as plasticizers for industrial, medical, and domestic purposes. Di(2-ethylhexyl) phthalate (DEHP) is one of the most synthetized PAEs and is considered resistant to the biological degradation

* Corresponding author address: Department of Civil and Environmental Engineering, Faculty of Science and Technology, University of Macau, Macau SAR, China. Email: hjshim@umac.mo.

due to its long hydrocarbon chain. An indigenous microorganism was isolated from the sludge collected from a local wastewater treatment plant (Macau SAR, China) to remove DEHP from artificially contaminated water. The 16S rRNA gene sequence analysis identified the microbial strain as *Acinetobacter* sp. SN13. Such major experimental parameters as pH (6-9) and temperature (35°C) were further optimized to improve the DEHP biodegradation efficiency. The growth kinetics followed the inhibition model (simulated using Matlab), with half saturation constant (272.3 mg l^{-1}), maximum degradation rate (124.8 mg l^{-1} day^{-1}), and inhibition constant (720.5 mg l^{-1}) estimated for the DEHP degradation, and half saturation constant (137.6 mg l^{-1}), specific growth rate (0.1192 day^{-1}), and inhibition constant (850.3 mg l^{-1}) for the microbial growth on DEHP. Since many environmental sites are contaminated with a mixture of inorganic and organic contaminants, the effect (inhibitory/stimulatory) of some microelements commonly present in wastewater (Fe^{3+} and Mn^{2+}) on DEHP biodegradation was also evaluated. The biodegradation performance of the isolate was improved as the Fe^{3+} concentration increased (100-1,000 μg l^{-1}), while higher Mn^{2+} concentrations (500-1,000 μg l^{-1}) inhibited the DEHP biodegradation. The aerobic biodegradation of phthalates generally occurs in two stages. First, phthalate diesters (PDEs) hydrolyze to phthalate monoesters (PMEs) followed by the PMEs hydrolysis to phthalic acid (PA) and then, the PA mineralization takes place by different mechanisms. For the Gram-negative bacteria like *Acinetobacter* sp., PA is usually further degraded via the dioxygenase-catalyzed pathways to protocatechuate (3,4-dihydroxy-benzoate) through 4,5-dihydroxyphthalate and *cis*-4,5-dihydroxy-4,5-dihydrophthalate. The respective DEHP degradation pathway for *Acinetobacter* sp. SN13 is proposed through the identification of mono-(2-ethylhexyl) phthalate (MEHP), PA, 3-katoadipate, β-carboxy-*cis,cis*-muconic acid, and protocatechuate by LC-MS.

Keywords: *Acinetobacter* sp., biodegradation pathway, di(2-ethylhexyl) phthalate, growth kinetics, microelements

INTRODUCTION

Phthalate esters (PAEs) are industrial chemicals widely used as plasticizers in industrial, domestic, and medical applications. PAEs belong to the emerging contaminants and had been considered as endocrine disrupting chemicals (EDCs) due to their adverse effects on fertility of humans and aquatic organisms (Fromme et al. 2002; Liu et al. 2014; Wang et al. 2015). PAEs can be easily released into the environment because they are physically, rather than chemically, bonded to the plastic products matrices.

Di(2-ethylhexyl) phthalate (DEHP) is one of the most common plasticizers in production worldwide. In the European Union, DEHP accounts for one-third of the phthalates manufactured while in China the production can reach 80% (Gao and Wen, 2016). In China, DEHP has been detected at very high levels in sediments, rivers, and wastewaters (He et al. 2013; Wang et al. 2014; Li et al. 2015). Wang et al. (2011) reported the average concentration of DEHP in industrial wastewater, well, and pond water was 135.68, 42.43, 14.20 $\mu g\ l^{-1}$, respectively, all over the standard value (8 $\mu g\ l^{-1}$) for drinking water and ambient surface water in China. In addition, DEHP is considered one of the most resistant PAEs due to its long hydrocarbon chain (Chang et al. 2004).

DEHP can be removed from the contaminated environments by physicochemical and biological methods (Magdouli et al. 2013). Biological technologies are usually carried out under aerobic and anaerobic conditions (Chao and Cheng, 2007; Chen et al. 2007b; Meng et al. 2015) while physicochemical methods include adsorption and advanced oxidation processes (AOPs) (Comninellis et al. 2008; Chen et al. 2009; Espinoza et al. 2016). Generally, AOPs such as photocatalytic oxidation, O_3/H_2O_2, and electro-oxidation show better performances on the DEHP removal. However, these technologies have some drawbacks that limit their application toward the DEHP removal at wastewater treatment plants. AOPs are considered not appropriate to be used when the high concentration of DEHP is present in wastewater treatment plant effluent and the performance can also be affected by the presence of other organic compounds (Chen et al. 2007a). In addition, AOPs can generate harmful by-products and require higher chemical consumption and relatively high treatment cost (Venkata Mohan et al. 2007). Compared to the physicochemical methods, biological processes are considered more applicable and environmental friendly and less costly (Zolfaghari et al. 2014).

The aim of this research was to evaluate the DEHP degradation performance using an indigenous microorganism isolated from a local wastewater treatment plant. Some major experimental parameters (temperature and pH) were further optimized to enhance the biodegradation efficiency for DEHP. The effect (stimulatory/inhibitory) of microelements (Fe^{3+} and Mn^{2+}) at different concentration levels (100, 500, and 1,000 µg l^{-1}) on DEHP biodegradation was also assessed for the microbial activity since many environmental sites are contaminated with organics simultaneously with inorganics. The biodegradation kinetics as well as the DEHP metabolic products were also studied/identified accordingly.

MATERIALS AND METHODS

Chemicals

DEHP (99% purity) was purchased from Sigma-Aldrich (U.S.A.). The DEHP stock solution (10 g l^{-1}) was prepared in dimethylformamide (DMF) due to its low water solubility (3 µg l^{-1}; JRC European Commission, 2008) and stored at 4°C. Manganese sulfate ($MnSO_4$) and ferric sulfate [$Fe_2(SO_4)_3$] were prepared at 0.275 g l^{-1} and 0.358 g l^{-1}, respectively, resulting in 0.1 g l^{-1} as Mn^{2+} and Fe^{3+} ions. All the other chemicals used were of analytical grades.

Culture Media

Three kinds of microbial culture media were used. First was nutrient broth (NB) consisting of 5.0 g peptone and 3.0 g beef extract (in 1 l deionized water). NB is a rich medium which supports the growth of a variety of microorganisms (Acumedia, 2010). This medium was used to cultivate all the cultivatable indigenous microorganisms present in the sludge sample. The second medium was the basal salt medium (BSM) which contained (in g l^{-1}) NaCl 1.0, K_2HPO_4 1.0, NH_4Cl 0.5, and $MgSO_4$ 0.4. This medium was used to subculture microorganisms capable of utilizing DEHP as carbon and energy source. The increasing DEHP concentrations were provided on a weekly basis (10-500 mg l^{-1}) in 160-ml serum bottles. The third was the nutrient broth agar consisting of 5.0 g peptone, 3.0 g beef extract, and 15 g agar per liter of water. The pH of the medium was adjusted to 7.0 by adding either NaOH or HCl (0.1 mol l^{-1}). All the apparatus and media were autoclaved for 20 min at 121°C under 103.5 kPa in advance.

Microbial Enrichment and Isolation

Before carrying out the enrichment experiments by using DEHP as the sole carbon source, pre-experiments were designed by inoculating microorganisms into BSM with DMF (1 ml) and Tween 80 (0.1 ml) separately to find out whether the organic solvent and the surfactant could support the microbial growth. After 5 days of incubation, no increase in the optical density was observed, confirming both organic compounds could not support the microbial growth. Tween 80 was used to ensure the homogeneity of DEHP in solution. The concentration applied (2.4 g l^{-1}) was higher than the critical micellar concentration (CMC) of this surfactant (15 mg l^{-1}; Chou et al. 2005). In addition, Tween 80 was not toxic to the isolate at the concentration used.

The activated sludge samples used to enrich and isolate the microbial strain were collected from a local wastewater treatment plant in Macau Special Administrative Region (SAR), China. Five grams of sludge samples were first added into the NB in 160 ml-serum bottles, followed by the addition of 10 mg l^{-1} DEHP as a co-substrate. The bottles were covered with a stopper, sealed with aluminum crimp, and placed on an orbital shaker at 25°C and 150 rpm. Five milliliters of inocula from the bottles were further inoculated into the BSM containing 50 mg l^{-1} DEHP as a sole growth substrate. The subsequent subculturing was performed by adding 5 ml of inocula, with the increasing DEHP concentration from 50 mg l^{-1} to 500 mg l^{-1} on a weekly basis. After several weeks, one pure culture was isolated from the nutrient agar (NA) plates.

Microbial Identification

The isolated microorganism with a higher ability to degrade DEHP was identified using the 16S rRNA gene sequence analysis. The detailed information about the microbial identification can be found elsewhere (Xu et al. 2017). The 16S rDNA sequences of the isolated strain have been recorded in the NCBI GenBank under the accession number KX_670538.

Experimental Setup

Two major experimental parameters (temperatures at 25, 30, 35°C and pHs at 3-9) were studied and optimized to improve the DEHP biodegradation. Five milliliters of inoculum were added into the serum bottles containing 45 ml BSM and spiked with 100 mg l^{-1} DEHP. The bottles were covered with stoppers (90% Teflon/10% silicone) and sealed with aluminum crimp. Controls were prepared under the same experimental conditions but without the inoculum. Samples were analyzed daily during 5 days of incubation.

For the kinetics study, the experimental procedure used was the same as described above except with different initial DEHP concentrations (10-500 mg l^{-1}) at pH 7 ± 0.2 and 30°C. The concentrations of DEHP and the optical density at 600 nm (OD_{600}) were determined every 24 h during 5 days. The effects of two microelements (Mn^{2+} and Fe^{3+}) on DEHP biodegradation were studied in the bottles containing 45 ml BSM, 5 ml inoculum, and DEHP at 100 mg l^{-1}. The bottles were individually spiked with the respective microelements at different concentrations (100, 500, 1,000 μg l^{-1}). The bottles without microelements were set as controls.

All the treatments were incubated in the dark (30ºC and 150 rpm), in duplicates, and each bottle was measured twice for the DEHP concentration. The one-way analysis of variance (ANOVA) was used to determine any significant differences during the temperature and pH optimization.

Analytical Assay

Samples from the serum bottles were collected daily and filtered through 0.45 μm membrane filters (Millipore®) before the high performance liquid chromatography (HPLC) analyses. The DEHP concentration was measured following Shailaja et al. (2008) using a Dionex UltiMate 3000 HPLC equipped with AcclaimTM C18 reversed-phase column (5 μm, 4.6 x 150 mm) and Dionex UltiMate 3000 diode array detector. The injection volume was 20 μl and the ratio of mobile phase (acetonitrile: deionized water) was 9:1. The flow rate of eluent was 0.5 ml min^{-1} and the oven temperature was 45°C. The linearity of the calibration curve was assessed over the range from 10 to 500 mg l^{-1} (DEHP), in triplicates, by the external standard method with a correlation coefficient equal to 0.9977. The precision of the analytical method was checked by the relative standard deviation (RSD) (using 100 mg l^{-1} DEHP standard solution) in two levels: repeatability (intra-assay precision, on same day, n=6) and intermediate precision (inter-assay precision, on different days, d=6). The RSD values for repeatability and intermediate precision were 0.51% and 1.35%, respectively. The accuracy (100.4±1.67%) was checked through a spike-recovery assessment of DEHP (100 mg l^{-1}) in basal salt medium. The precision and accuracy of the analytical method were within the acceptable limits.

The intermediate metabolites were analyzed by liquid chromatography-mass spectrometry (LC-MS) with the electro spray ionization (ESI) source, using a full scan for the mass range of 50 to 400.

The ESI mass spectral data were obtained in positive and negative scan modes and the probe temperature was 300°C.

The magnitude of OD was measured spectrophotometrically (Shimadzu, Japan) at a wavelength of 600 nm.

RESULTS AND DISCUSSION

Microbial Identification

The isolate (*Acinetobacter* sp. SN13) is shown belonging to the family of Moraxellaceae by analyzing the phylogenetic tree (Figure 1). The microorganism is a Gram-negative coccal shape bacterium and most related (more than 98% identical) to *Acinetobacter genomosp. Acinetobacter* species commonly exist in soils and waters and can survive in both dry and moist surfaces (Xu et al. 2017). This microorganism is able to grow at a wide range of temperature conditions. The isolate belongs to the same genus as the one isolated by Latorre et al. (2012), which was the first *Acinetobacter* strain isolated from the landfill leachate with the capability of using DEHP as carbon and energy source.

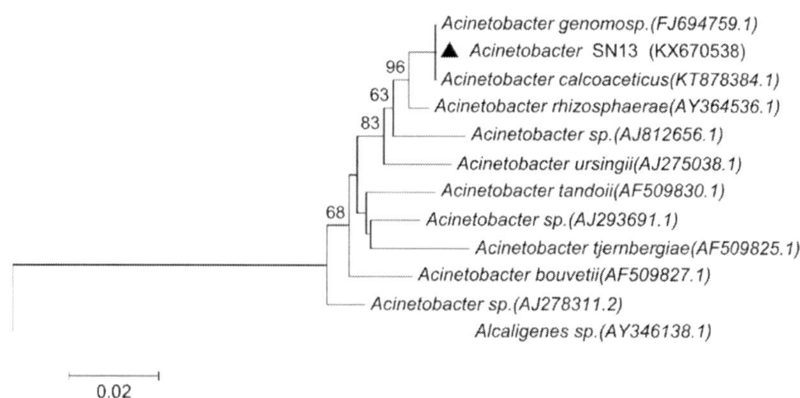

Figure 1. Neighbor-joining trees based on the sequences of 16S rRNA gene numbers.

Effects of Temperature and pH on DEHP Biodegradation

The DEHP concentration sharply decreased to 60 mg l^{-1} within one day at 35°C and 30°C, whereas it was still almost 90 mg l^{-1} at 25°C. The biodegradation efficiencies at 35°C and 30°C were 10.32 and 6.88% higher than at 25°C. However, there was no significant difference between 35°C and 30°C ($p = 0.22$). Therefore, 30°C was used in further experiments. Figure 2 shows the DEHP removal together with the microbial growth under different pH conditions. Under the acidic condition, both DEHP removal and microbial growth were much lower compared to neutral or alkaline conditions. According to Fang et al. (2010), the enzymes activities involved in the PAEs degradation are sensitive to low pHs and this would further explain the low DEHP removal efficiency at the acidic pH. Dong et al. (2015) and Gururaj et al. (2016) also studied the activity of *Acinetobacter* strain at pH 3 to 10, with similar results. The isolate showed the highest activity at pH 8 and when pH was over 10 or less than 6, the activity decreased rapidly, suggesting it is more adapted to neutral and weak alkaline conditions. The removal efficiencies for DEHP at pHs 6-9 showed no significant differences ($p = 0.87$). Similar results were obtained by others on the PAEs bioremoval (Chen et al. 2007b; Jin et al. 2010; Zhang et al. 2014; Meng et al. 2015).

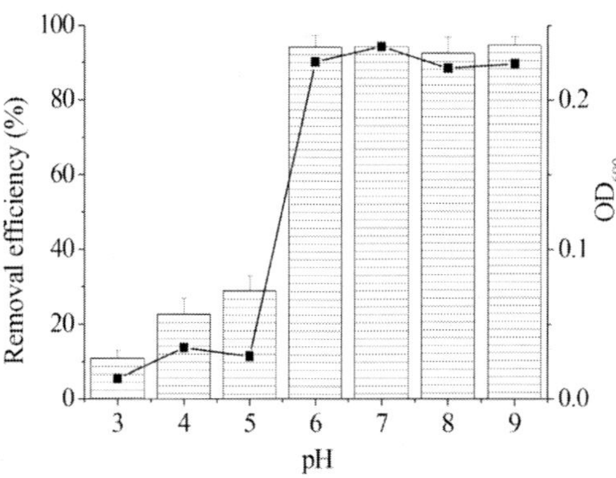

Figure 2. DEHP removal efficiency (bar) by *Acinetobacter* sp. and microbial growth (OD line) at different pHs (3-9).

Biodegradation Kinetics

Figure 3 shows the cell growth (OD_{600}) at different DEHP initial concentrations (100-500 mg l^{-1}). A lag phase was observed at 500 mg l^{-1}, suggesting the DEHP concentration at ≥500 mg/L inhibitory to the isolate growth. The result is in agreement with Lu et al. (2009) and Meng et al. (2015) who also reported no lag phase for the PAEs biodegradation observed within the same concentration level. The inhibitory concentration levels of DEHP also varied with microorganisms. The growth of *Pseudoxanthomonas* sp. strain N4 was affected when the DEHP concentration was higher than 500 mg l^{-1} (Meng et al. 2015) while for *Microbacterium sp.* strain CQ0110Y, the growth was inhibited when the DEHP concentration was higher than 1,600 mg l^{-1} (Chen et al. 2007b).

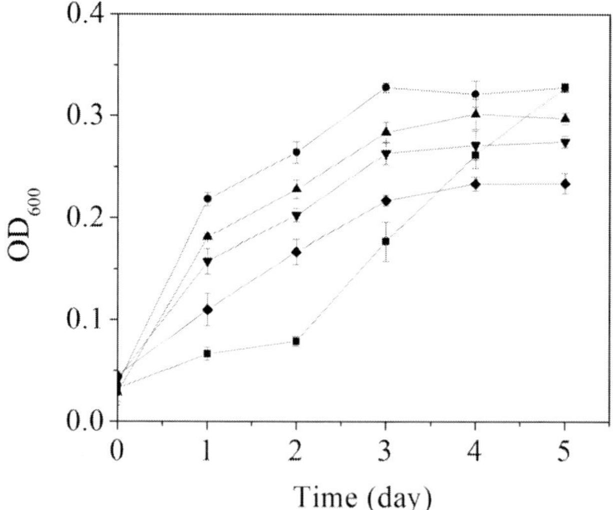

Figure 3. Cell growth of *Acinetobacter* sp. at different DEHP concentrations: (■) 500; (●) 400; (▲) 300; (▼) 200; (◆) 100 mg l^{-1}.

Since the substrate was inhibitory to the isolate at concentrations higher than 400 mg l^{-1}, the kinetics for the DEHP degradation rate (Eq. 1) and the specific growth rate (Eq. 2) were estimated as follows (Shim and Yang, 1999):

$$D = \frac{D_m S}{K_S + S + S^2 / K_i} \quad (1)$$

$$\mu = \frac{\mu_m S}{K_S + S + S^2 / K_i} \quad (2)$$

where D (mg l^{-1} day^{-1}) and D_m (mg l^{-1} day^{-1}) are degradation rate and maximum degradation rate, respectively, S (mg l^{-1}) is the substrate (DEHP) concentration, K_s (mg l^{-1}) is the half-saturation constant, K_i (mg l^{-1}) is the inhibition constant, μ is the specific growth rate, and μ_m is the maximum specific growth rate. The inhibition models for both degradation rate and specific growth rate was further simulated using the Matlab software. The comparison between simulated and experimentally obtained values is shown in Figures 4 (a) and (b).

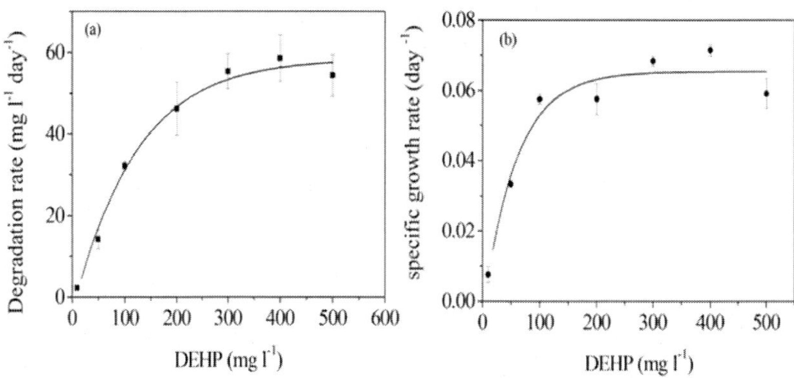

Figure 4. (a) Degradation kinetics and (b) mnicrobial growth kinetics for DEHP. (■) and (●) are the experimental data and (—) is the simulated by Matlab.

The maximum degradation rate (D_m) and maximum specific growth rate (μ_m) were 124.8 mg l^{-1} day^{-1} and 0.1192 day^{-1}, respectively. The half-saturation (K_s) and inhibition (K_i) constants for the degradation kinetics were 272.3 mg l^{-1} and 720.5 mg l^{-1} while for the microbial growth kinetics, both constants were 137.6 mg l^{-1} and 850.3 mg l^{-1}, respectively. The inhibitory effect of high DEHP concentrations on microbial growth is probably associated with the DEHP toxicity. However, it is also possible DEHP could present the inhibitory effect on growth even at lower concentrations because at low concentrations of DEHP as carbon and energy source, this compound could be considered a growth limiting factor, leading to the reduction of microbial growth to some extent (Xu et al. 2017). Table 1 compares the specific degradation rates of DEHP for different microorganisms. *Acinetobacter* sp. in current study shows a higher specific degradation rate compared to most other microorganisms (Chang et al. 2004; Chen et al. 2007b; Meng et al. 2015), further indicating this isolate with a better ability for the DEHP degradation as a promising alternative for the biological removal of DEHP from contaminated sites. Considering DEHP is one of the highest molecular weight phthalates, the developed biological process could also be applied to other phthalates, including the ones with lower molecular weights.

Table 1. Comparison of specific degradation rates for DEHP by different microorganisms

Microorganism	**Initial concentration (mg l^{-1})**	**Specific degradation rate (x 10^{-7} mg day^{-1} $cell^{-1}$)**	**Reference**
Corynebacterium sp. DK4	100	1.42	Chang et al. (2004)
Microbacterium sp. CQ0110Y	2,000	4.71	Chen et al. (2007b)
Pseudoxanthomonas sp.	500	1.60	Meng et al. (2015)
Achromobacter denitrificans SP1	3,900	27.37	Pradeep et al. (2015a)
***Acinetobacter* sp.**	**400**	**21.71**	**This study**

Effects of Fe^{3+} and Mn^{2+} on DEHP Biodegradation

Figures 5 (a) and (b) show the bioremoval efficiencies for DEHP after the addition of Fe^{3+} and Mn^{2+} at different concentration levels (100, 500, and 1,000 μg l^{-1}).

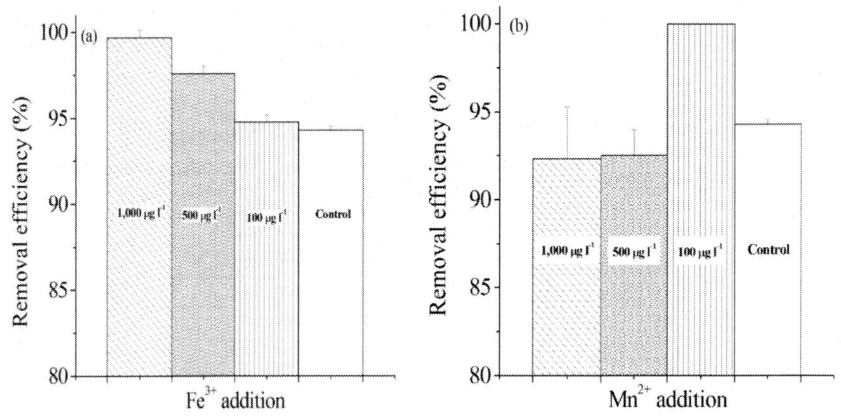

Figure 5. Effects of (a) Fe^{3+} and (b) Mn^{2+} addition on DEHP removal efficiency.

The DEHP removal efficiencies were higher in the presence of Fe^{3+}, regardless of the concentration tested. Dioxygenase is a key enzyme produced by gram-negative bacteria to catalyze the PA cleavage during DEHP biodegradation under the aerobic condition (Benjamin et al. 2015) and the Fe^{3+} ion plays an important role in the dioxygenase composition (Di Nardo et al. 2004; Karimpour et al. 2013). The presence of Fe^{3+} can also stimulate the production of enzymes in microorganisms associated with the aerobic biodegradation of hydrocarbons (Wei and Chu, 1998). The addition of Fe^{3+} (0.1 mM) improved both cell density and degradation efficiency for anthracene for *Pseudomonas* sp. (Santos et al. 2008). One reason for the DEHP biodegradation improvement in this study in the presence of Fe^{3+} could be associated with the stimulation of enzyme production in *Acinetobacter* sp. The increased concentration of Fe^{3+} as microelement may stimulate the production of dioxygenases to catalyze the cleavage of protocatechuate, one of the key intermediates in the DEHP biodegradation. This cleavage is catalyzed by 3,4-dioxygenase and metal ions such as Fe^{3+} can influence the enzyme activity due to their specific mechanism and nature. On the other hand, 100 μg l^{-1} Mn^{2+} showed the highest removal efficiency for DEHP while there was no significant difference between 1,000 and 500 μg l^{-1} Mn^{2+} and the control without Mn^{2+}. Some studies have also found the metal ions such as Mn^{2+} at low concentration can enhance the fungal biodegradation ability by stimulating the enzymes production and the activity of enzyme as inducer (Song et al. 2013; Knežević et al. 2014). The manganese ion (Mn^{2+}) has been used as an inducer of manganese peroxidase (MnP) activity for *Phanerochaete chrysosporium*, *Dichomitus squalens*, and *Phlebia radiate* (Yanto and Tachibana, 2014). Song et al. (2013) found the supplement of manganese can improve the lignin degradation ability of *Irpex lacteus*. However, very few studies focus on the effect of Mn^{2+} on the enzyme activity of bacteria. Adams and Ghiorse (1985) reported the high concentrations of Mn^{2+} decreased the cell yield due to the toxicity. As shown in Figure 5 (b), the presence of Mn^{2+} at the lower concentration (100 μg l^{-1}) increased the removal efficiency for DEHP, suggesting Mn^{2+} can stimulate the enzyme production for the DEHP degradation at low concentrations, while inhibition of cell growth and enzyme activity was observed at higher Mn^{2+} concentrations. Since the DEHP biodegradation pathway is complex and involves different types of enzymes, higher concentrations of manganese ion used in this study may affect/deactivate the enzymes to different extent. To confirm whether the inhibitory/stimulatory effects observed in current study were due to the presence of cations (Fe^{3+} and Mn^{2+}) or anion (SO_4^{2-}), different anions such as chloride using $FeCl_3$ and $MnCl_2$ are to be tested.

DEHP Biodegradation Pathway

The aerobic phthalates biodegradation pathways consist of two stages: the hydrolysis of phthalate diesters (PDEs) to phthalate monoesters (PMEs) and then to phthalic acid (PA), followed by the mineralization of PA (Liang et al. 2008; Benjamin et al. 2015; Li et al. 2016). Roslev et al.

(1998) reported the DEHP degradation involves several enzymes, including hydrolases, dehydrogenases, decarboxylases, and oxygenases. *Acinetobacter* species are known to produce lipolytic enzymes (esterases and lipases) (Breuil and Kushner, 1975; Shabtai and Gutnick, 1985; Kim et al. 2003), for the hydrolysis of ester bonds (Rao et al. 2013). The first step of DEHP degradation should be the hydrolysis of one ester bond to generate the respective acid mono-(2-ethylhexyl) phthalate (MEHP) and alcohol (2-ethylhexanol) (Magdouli et al. 2013; Pradeep et al. 2015b). In current study, MEHP was identified as one of intermediates (m/z=277) and its chromatographic peak area increased during the first 3 days of incubation. The other intermediate, 2-ethylhexanol, was not identified during 5 days of incubation. Instead, its oxidation product, di-ethyl hexanoic acid, was identified (m/z = 145), which confirms the microbial oxidation of primary alcohol to carboxylic acid. *Acinetobacter* spp. also produce the alcohol dehydrogenases to oxidize alcohols all the way to carboxylic acids (Singer and Finnerty, 1985). These reactions are probably involved in the early stages of DEHP biodegradation. In the next step, MEHP can be further hydrolyzed to PA. Roslev et al. (1998) reported PA was degraded much faster by the indigenous microorganisms than DEHP, meaning as soon as PA was produced from MEHP through hydrolysis, it would be further metabolized into other intermediates. This might explain why PA was not detected throughout the incubation period, further suggesting the isolate could effectively use PA as carbon and energy source. According to Wu et al. (2013), PA can be further metabolized through two well-established pathways, for both Gram-negative and Gram-positive bacteria. For the Gram-negative including the isolate *Acinetobacter* sp., it has been reported that PA can be further transformed to protocatechuate (3,4-dihydroxy-benzoate) through *cis*-4,5-dihydroxy-4,5-dihydrophthalate and 4,5-dihydroxyphthalate via the dioxygenase-catalyzed pathways (Liang et al. 2008; Wu et al. 2013). Other intermediates identified were β-carboxy-*cis*,*cis*-muconic acid (m/z=186) and 3-katoadipate (m/z=79), which were generated via the *ortho*-clevage pathway for protocatechuate. According to BioCyC database collection (2016), the fission of the protocatechuate aromatic ring can be catalyzed by one of three distinct dioxygenases, protocatechuate 4,5-dioxygenase, protocatechuate 3,4-dioxygenase, and protocatechuate 2,3-dioxygenase. The intermediate 3-katoadipate is then oxidized all the way to acetyl-CoA which enters the tricarboxylic acid cycle (TCA) and is mineralized to carbon dioxide and water (Chatterjee and Dutta, 2008). As shown in Figure 6, the proposed DEHP degradation pathway for the isolate *Acinetobacter* sp. is in accordance with BioCyC database collection (2016).

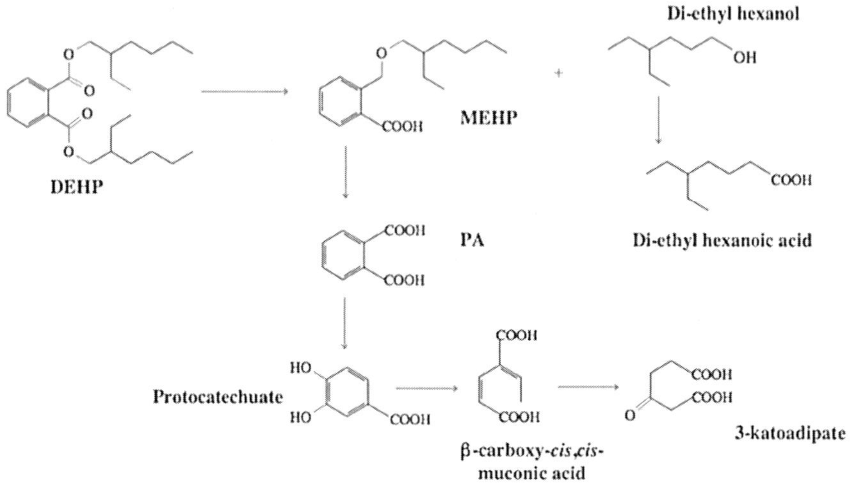

Figure 6. The proposed DEHP degradation pathway for *Acinetobacter* sp. SN13.

CONCLUSION

An indigenous bacterial isolate, enriched from the activated sludge and identified as *Acinetobacter* sp. SN13, could degrade DEHP very efficiently. The optimal temperature and pH were 30°C and neutral or alkaline condition, respectively. The isolate showed the inhibition kinetics for both DEHP degradation and cell growth, at 500 mg l^{-1} DEHP or higher. The Fe^{3+} ion at 100-1,000 μg l^{-1} showed stimulatory to the DEHP biodegradation, while Mn^{2+} was stimulatory at the lower concentration (100 μg l^{-1}) but inhibitory at higher concentrations (500-1,000 μg l^{-1}). The

DEHP degradation pathway for the isolate is proposed with some metabolic products identified. The DEHP degradation performance by this isolate is much better compared to previous studies using other microorganisms. In addition, this is the first time to study the association between PAEs biodegradation and microelements, which can be helpful and significant when this biological process applied in practice. The biological process developed could be further scaled up by using an appropriate bioreactor configuration and applied to treat different types of wastewater continuously, especially the ones containing high concentration levels of DEHP and other PAEs generated from the plastics manufacturing industries.

ACKNOWLEDGMENTS

This research was supported by the University of Macau Multi-Year Research Grant (MYRG2014-00112-FST) and by grants from the Macau Science and Technology Development Fund (FDCT/061/2013/A2 and FDCT/063/2013/A2).

REFERENCES

Acumedia, 2010. Nutrient broth. http://foodsafety.neogen.com/pdf/Acumedia_PI/7146_PI.pdf (Accessed on 15 January 2017).

Adams L. F., Ghiorse W. C., 1985. Influence of manganese on growth of a sheathless strain of *Leptothrix discophora*. *Appl. Environ. Microbiol.* 49, 556-562.

Benjamin S., Pradeep S., Josh M. S., Kumar S., Masai E., 2015. A monograph on the remediation of hazardous phthalates. *J. Hazard. Mater.* 298, 58-72.

BioCyC database collection, 2016. *Acinetobacter baumannii* 6013150 Pathway: protocatechuate degradation II (*ortho*-cleavage pathway). <http://www.biocyc.org/ABAU525243-HMP/NEW-IMAGE?type=PATHWAY&object=PROTOCATECHUATE-ORTHO-CLEAVAGE-PWY> (Accessed on 16 February 2017).

Breuil C., Kushner D. J., 1975. Lipase and esterase formation by psychrophilic and mesophilic *Acinetobacter* species. *Can. J. Mirobiol.* 21, 423-433.

Chang B. V., Yang C. M., Cheng C. H., Yuan S. Y., 2004. Biodegradation of phthalate esters by two bacteria strains. *Chemosphere* 55, 533-538.

Chao W. L., Cheng C. Y., 2007. Effect of introduced phthalate-degrading bacteria on the diversity of indigenous bacterial communities during di-(2-ethylhexyl) phthalate (DEHP) degradation in a soil microcosm. *Chemosphere* 67, 482-488.

Chatterjee S., Dutta T. K., 2008. Complete degradation of butyl benzyl phthalate by a defined bacterial consortium: role of individual isolates in the assimilation pathway. *Chemosphere* 70, 933-941.

Chen C. Y., Chen C. C., Chung Y. C., 2007a. Removal of phthalate esters by acyclodextrin-linked chitosan bead. *Bioresour. Technol.* 98, 2578-2583.

Chen J. A., Li X., Li J., Cao J., Qiu Z., Qing Z., Shu W., 2007b. Degradation of environmental endocrine disruptor di-2-ethylhexyl phthalate by a newly discovered bacterium, *Microbacterium* sp. strain CQ0110Y. *Appl. Microbiol. Biotechnol.* 74, 676-682.

Chen C. Y., Wu P. S., Chung Y. C., 2009. Coupled biological and photo-Fenton pretreatment system for the removal of di-(2-ethylhexyl) phthalate (DEHP) from water. *Bioresour. Technol.* 100, 4531-4534.

Chou D. K., Krishnamurthy R., Randolph T. W., Carpenter J. F., Manning M. C., 2005. Effects of Tween 20 and Tween 80 on the stability of Albutropin during agitation. *J. Pharm. Sci.* 94(6), 1368-1381.

Comninellis C., Kapalka A., Malato S., Parsons S. A., Poulios I., Mantzavinos D., 2008. Advanced oxidation processes for water treatment: advances and trends for R&D. *J. Chem. Technol. Biotechnol.* 83, 769-776.

Di Nardo G., Tilli S., Pessione E., Cavaletto M., Giunta C., Briganti F., 2004. Structural roles of the active site iron (III) ions in catechol 1,2-dioxygenases and differential secondary structure changes in isoenzymes A and B from *Acinetobacter radioresistens* S13. *Arch. Biochem. Biophys.* 431, 79-87.

Dong W., Jiang S., Shi K., Wang F., Li S., Zhou J., Huang Y., 2015. Biodegradation of fenoxaprop-P-ethyl (FE) by *Acinetobacter* sp. strain DL-2 and cloning of FE hydrolase gene afeH. *Bioresour. Technol.* 186, 114-121.

Espinoza J. D. G., Drogui P., Zolfaghari M., Dirany A., Ledesma M. T. O., Gortáres-Moroyoqui P., Buelna G., 2016. Performance of electrochemical oxidation process for removal of di (2-ethylhexyl) phthalate. *Environ. Sci. Pollut. Res.* 23, 1-10.

Fang C. R., Yao J., Zheng Y. G., Jiang C. J., Hu L. F., Wu Y. Y., Shen D.S., 2010. Dibutyl phthalate degradation by *Enterobacter* sp. T5 isolated from municipal solid waste in landfill bioreactor. *Int. Biodeter. Biodegr.* 64, 442-446.

Fromme H., Küchler T., Otto T., Pilz K., Müller J., Wenzel A., 2002. Occurrence of phthalates and bisphenol A and F in the environment. *Water Res.* 36, 1429-1438.

Gao D. W., Wen Z. D., 2016. Phthalate esters in the environment: A critical review of their occurrence, biodegradation, and removal during wastewater treatment processes. *Sci. Total Environ.* 541, 986-1001.

Gururaj P., Ramalingam S., Devi G. N., Gautam P., 2016. Process optimization for production and purification of a thermostable, organic solvent tolerant lipase from *Acinetobacter* sp. AU07. *Braz. J. Microbiol.* 47, 647-657.

He W., Qin N., Kong X., Liu W., He Q., Xu F., 2013. Spatio-temporal distributions and the ecological and health risks of phthalate esters (PAEs) in the surface water of a large, shallow Chinese lake. *Sci. Total Environ.* 461, 672-680.

Jin D. C., Liang R. X., Dai Q. Y., Zhang R. Y., Wu X. L., Chao W. L., 2010. Biodegradation of di-n-butyl phthalate by *Rhodococcus* sp. JDC-11 and molecular detection of 3,4-phthalate dioxygenase gene. *J. Microbiol. Biotechnol.* 20, 1440-1445.

JRC European Comission, 2008. https://echa.europa.eu/documents/10162/060d4981-4dfb-4e40-8c69-6320c9debb01 (Accessed on 21 February 2017).

Karimpour T., Safaei E., Wojtczak A., Jagličić Z., Kozakiewicz A., 2013. Iron (III) complexes of ethylenediamine derivatives of aminophenol ligands as models for enzyme–substrate adducts of catechol dioxygenases. *Inorg. Chim. Acta* 395, 124-134.

Kim Y. H, Lee J, Moon S. H., 2003. Degradation of an endocrine disrupting chemical, DEHP [di-(2-ethylhexyl)-phthalate], by *Fusarium oxysporum* f. sp pisi cutinase. *Appl Microbiol. Biotechnol.* 63, 75-80.

Knežević A., Stajić M., Vukojević J., Milovanović I., 2014. The effect of trace elements on wheat straw degradation by *Trametes gibbosa*. *Int. Biodeter. Biodegrad.* 96, 152-156.

Latorre I., Hwang S., Montalvo-Rodriguez R., 2012. Isolation and molecular identification of landfill bacteria capable of growing on di-(2-ethylhexyl) phthalate and deteriorating PVC materials. *J. Environ. Sci. Health Part A* 47, 2254-2262.

Li B., Hu X., Liu R., Zeng P., Song Y., 2015. Occurrence and distribution of phthalic acid esters and phenols in Hun River Watersheds. *Environ. Earth Sci.* 73, 5095-5106.

Li D., Yan J., Wang L., Zhang Y., Liu D., Geng H., Xiong L., 2016. Characterization of the phthalate acid catabolic gene cluster in phthalate acid esters transforming bacterium-*Gordonia* sp. strain HS-NH1. *Int. Biodeter. Biodegrad.* 106, 34-40.

Liang D. W., Zhang T., Fang H. H., He J., 2008. Phthalates biodegradation in the environment. *Appl. Microbiol. Biotechnol.* 80, 183-198.

Liu X., Shi J., Bo T., Zhang H., Wu W., Chen Q., Zhan X., 2014. Occurrence of phthalic acid esters in source waters: a nationwide survey in China during the period of 2009–2012. *Environ. Pollut.* 184, 262-270.

Lu Y., Tang F., Wang Y., Zhao J., Zeng X., Luo Q., Wang L., 2009. Biodegradation of dimethyl phthalate, diethyl phthalate and di-n-butyl phthalate by *Rhodococcus* sp. L4 isolated from activated sludge. *J. Hazard. Mater.* 168, 938-943.

Magdouli S., Daghrir R., Brar S. K., Drogui P., Tyagi R. D., 2013. Di 2-ethylhexylphtalate in the aquatic and terrestrial environment: a critical review. *J. Environ. Manage.* 127, 36-49.

Meng X., Niu G., Yang W., Cao X., 2015. Di (2-ethylhexyl) phthalate biodegradation and denitrification by a *Pseudoxanthomonas* sp. strain. *Bioresour. Technol.* 180, 356-359.

Pradeep S., Josh M. S., Binod P., Devi R. S., Balachandran S., Anderson R. C., Benjamin S., 2015a. *Achromobacter denitrificans* strain SP1 efficiently remediates di (2-ethylhexyl) phthalate. *Ecotoxicol. Environ. Safety* 112, 114-121.

Pradeep S., Josh M. S., Hareesh E. S., Kumar S., Benjamin S., 2015b. *Achromobacter denitrificans* strain SP1 produces an intracellular esterase upon utilizing di (2ethylhexyl) phthalate. *Int. Biodeter. Biodegrad.* 105, 160-167.

Rao L., Xue Y., Zheng Y., Lu J. R., Ma Y., 2013. A Novel Alkaliphilic *Bacillus* Esterase Belongs to the 13th Bacterial Lipolytic Enzyme Family. *PLoS ONE* 8(4) PMC3618048.

Roslev P., Madsen P. L., Thyme J. B., Henriksen K., 1998. Degradation of phthalate and di-(2-ethylhexyl) phthalate by indigenous and inoculated microorganisms in sludge-amended soil. *Appl. Environ. Microbiol.* 64, 4711-4719.

Santos E. C., Jacques R. J., Bento F. M., Maria do Carmo R. P., Selbach P. A., Sá E. L., Camargo F. A., 2008. Anthracene biodegradation and surface activity by an iron-stimulated *Pseudomonas* sp. *Bioresour. Technol.* 99, 2644-2649.

Shabtai Y., Gutnick D. L., 1985. Exocellular esterase and emulsan release from the cell surface of *Acinetobacter calcoaceticus* RAG-1. *J. Bacteriol.* 161, 1176-1181.

Shailaja S., Venkata Mohan S., Rama Krishna M., Sarma P. N., 2008. Degradation of di-ethylhexyl phthalate (DEHP) in bioslurry phase. *Int. Biodeter. Biodegrad.* 62. 143-152.

Shim H., Yang S. T., 1999. Biodegradation of benzene, toluene, ethylbenzene, and *o*-xylene by a coculture of *Pseudomonas putida* and *Pseudomonas fluorescens* immobilized in a fibrous-bed bioreactor. *J. Biotechnol.* 67, 99-112.

Singer M. E., Finnerty W. R., 1985. Alcohol Dehydrogenases in *Acinetobacter* sp. Strain HO1-N: Role in Hexadecane and Hexadecanol Metabolism. *J. Bacteriol.* 164(3), 1017-1024.

Song L., Ma F., Zeng, Y., Zhang X., Yu H., 2013. The promoting effects of manganese on biological pretreatment with *Irpex lacteus* and enzymatic hydrolysis of corn stover. *Bioresour. Technol.* 135, 89-92.

Venkata Mohan S., Shailaja S., Rama Krishna M., Sarma P., 2007. Adsorptive removal of phthalate ester (di-ethyl phthalate) from aqueous phase by activated carbon: a kinetic study. *J. Hazard. Mater.* 146, 278-282.

Wang Q., Wang L., Chen X., Rao K. M., Lu S. Y., Ma S. T., Wang J. S., 2011. Increased urinary 8-hydroxy-2'-deoxyguanosine levels in workers exposed to di-(2-ethylhexyl) phthalate in a waste plastic recycling site in China. *Environ. Sci. Pollut. Res.* 18, 987-996.

Wang J., Bo L., Li L., Wang D., Chen G., Christie P., Teng Y., 2014. Occurrence of phthalate esters in river sediments in areas with different land use patterns. *Sci. Total Environ.* 500, 113-119.

Wang W. L., Wu Q. Y., Wang C., He T., Hu H. Y., 2015. Health risk assessment of phthalate esters (PAEs) in drinking water sources of China. *Environ. Sci. Pollut. Res.* 22, 3620-3630.

Wei Y. H., Chu I. M., 1998. Enhancement of surfactin production in iron-enriched media by *Bacillus subtilis* ATCC 21332. *Enz. Microb. Technol.* 22, 724-728.

Wu J., Liao X., Yu F., Wei Z., Yang L., 2013. Cloning of a dibutyl phthalate hydrolase gene from *Acinetobacter* sp. strain M673 and functional analysis of its expression product in *Escherichia coli*. *Appl. Microbiol. Biotechnol.* 97, 2483-2491.

Xu J., Lu Q., Toledo R. A., Shim H., 2017. Degradation of di-2-ethylhexyl phthalate (DEHP) by an indigenous isolate *Acinetobacter* sp. SN13. *Int. Biodeter. Biodegr.* 117, 205-214.

Yanto D. H. Y., Tachibana S., 2014. Enhanced biodegradation of asphalt in the presence of Tween surfactants, Mn^{2+} and H_2O_2 by *Pestalotiopsis* sp. in liquid medium and soil. *Chemosphere* 103, 105-113.

Zhang C., Zeng G., Huang D., Lai C., Huang C., Li N., He X., 2014. Combined removal of di (2-ethylhexyl) phthalate (DEHP) and Pb(ii) by using a cutinase loaded nanoporous gold-polyethyleneimine adsorbent. *RSC Adv.* 4, 55511-55518.

Zolfaghari M., Drogui P., Seyhi B., Brar S. K., Buelna G., Dubé R., 2014. Occurrence, fate and effects of di (2-ethylhexyl) phthalate in wastewater treatment plants: A review. *Environ. Pollut.* 194, 281-293.

In: Micropollutants
Editor: Tabitha N. Holloway

ISBN: 978-1-53612-067-7

Chapter 3

THE LANTHANIDES AND PLATINUM GROUP METALS AS MINERAL MICROPOLLUTANTS IN RUSSIAN SOIL

Yu. N. Vodyanitskii[1,*], D. V. Ladonin[1] and A. T. Savichev[2,3]

[1]Lomonosov Moscow State University, Faculty of Soil Science, Moscow, Russia

[2]Russian Academy of Sciences' Geological Institute, Moscow, Russia

[3]Dokuchaev Soil Science Institute, Moscow, Russia

ABSTRACT

Pollution risk of soil contamination by the lanthanides (Ln) and platinum group metals (PGM) increases with industrialization and urbanization growth in Russia. The development of electronic, oil chemistry, metallurgy, medical industries rises the input of Ln-containing wastes into the soil. When considering soil contamination by Ln and PGM, it appears important to point out the way: basically, aerial,

---

* Corresponding author E-mail: yu.vodyan@mail.ru.

although hydrogenic contamination of alluvial soils is observed in some places too. Increasing the number of combustion engine with catalytic exhaust neutralizers, led to air and soil pollution by cancerogenic PGM: rhodium, palladium, platinum. In Russia, road dust analysis is widely used in the study of aerial soil contamination by Ln and PGM. Soils in Russia are contaminated by Ln and PGM mostly by aerial way near steel plants, thermal power coal-fired plants, as well as in large cities with tight traffic. Alluvial soils along rivers are contaminated hydrogenically by untreated industrial wastewater or by leachating from dumps, where waste is stockpiled after non-ferrous metals ores enrichment. In hydrogenically contaminated alluvial soils micropollutants are strongly accumulated in Fe-Mn-nodules, due to this eliminating from the geochemical cycle. The peat soils in the Western Siberia are contaminated by the lanthanides in the sites of oil spill.

Keywords: lanthanides, platinum grope metals, mineral micropollutants, aerial and hydrogenic contamination of soils, oil-contaminated soils

INTRODUCTION

14 elements, (lanthanum, cerium, praseodymium, neodymium, samarium, europium, gadolinium, terbium, dysprosium, holmium, erbium, thulium, ytterbium and lutetium) make up the group of the lanthanides (Ln). Together with yttrium the lanthanides refer to the group of rare earth elements. Due to special chemical properties, their content and way of accumulation in soils, the lanthanides are usually considered as a separate group of metals. Conditionally, the lanthanides are subdivided into light (La, Ce, Pr, Nd, Sm, Eu) and heavy (Gd, Tb, Dy, Ho, Er, Tm, Yu, Lu).

At present, the lanthanides' use is growing increasingly. Ore mining, metal and metal-products are a powerful source of soil pollution with these elements, as well as the production of mineral phosphate fertilizers, rare earth and ferrous metals (for example, Ce is used by doping ductile iron and iron alloys), and burning of coal with high natural content of the lanthanides in thermal power plants (Ivanov, 1997).

Another group of mineral micropollutants is represented by platinum group metals (PGM): ruthenium (Ru), rhodium (Rh), palladium (Pd), osmium (Os), iridium (Ir) and platinum (Pt). Currently, PGM are very important commercially. This led to environmental pollution and accumulation of PGM in the biosphere in an amount exceeded by several orders their natural background content. Platinum group metals are generally found in deposits of nickel, copper and iron ore (Ravindra et al. 2004), therefore, can enter the soil with emissions of metallurgical enterprises.

However, these sources of soil pollution with PGM are not major today. Now the main source of PGM pollution is motor transport. Platinum, palladium and rhodium are the part of the catalytic exhaust neutrolizers, so their content in soils rises rapidly.

1. Methods for Determination of the Lanthanides and Platinum Group Metals in Soils

The content of platinum group metals and the lanthanides are determined mainly by physical methods: regular neutron activation, ICP and specially modified XRF for these scattered heavy elements.

1.1. Activation Methods of Analysis (AMA)

Activation methods of analysis (AMA) are based on the irradiation of studied material by nuclear radiation and further measuring of the activity of radionuclides, formed by irradiation in nuclear reactions. Despite the wide variation of activation methods, the irradiation by thermal neutrons with energies of a range of 0.005-0.4 eV is most applyied (Kuznetsov, 1978). This is due to the high section of activation of a large number of elements, formation of radionuclides in nuclear reactions of almost the same type (n, γ), that reduced the number of confounding factors and made possible to use nuclear reactors as a powerful source of thermal neutron.

Samples, activated in the reactor, are measured in gamma-spectrometer settings after certain exposure cycles. Under irradiation of most chemical elements by thermal neutrons, radiative capture reaction (n, γ) takes place to form a radioactive nuclide. Resulting neutron-rich radioactive nuclides is usually decayed accompanied by the emission of beta-radiation and concomitant gamma-radiation.

AMA advantages are low limit of determinaion and simple preparation of samples. This is a non-destructive method. AMA disadvantage is a long period to complete analysis cycle - a month or more. Measurements sensitivity strongly depends on nuclear constants of investigated elements isotopes (neutron capture cross section, half-life, and so on), and is considerably different for various chemical elements. A significant disadvantage of this analysis is its high cost. Finally, there is a group of heavy metals, difficult to be determined by this method (Pr, Gd, Dy, Tm).

1.2. Inductively Coupled Plasma with Mass Spectrometry

Mass spectrometry method with inductively coupled plasma (ICP-MS) was developed in the 70s of XX century, when the ion source (at first - DC electric arc, and later - the plasma torch) has been linked with already existing quadrupole mass filter in united device through the special vacuum interface. Alan Gray published his work in 1975, where the first mass spectrum, obtained by ionization of the sample in the plasma, was shown (Gray, 1975). The first commercial plasma mass spectrometer was released in 1983. In the late 90-ies of the XX century, it became possible to use the ICP-MS method for the reliable analysis of environmental objects of complex matrix composition like soils, rocks, sediments, and so on.

This is due to certain requirements for equipment and sample preparation, which, in general, has been solved. Widespread of ICP-MS method is limited by high cost of equipment and supplies, increased demands on the purity of reagents, gas, water, clean rooms and so on.

In research we used Agilent Technologies 7500 Series ICP-MS.

1.3. X-Ray Fluorescence Analysis (XRF)

For the lanthanides analysis, this method had to be modified, as only yttrium Y from all rare earth elements can be determined by characteristic K-lines with excitation radiation from any X-ray tube.

Significantly expanding of the list of determined by K-lines lanthanides was done due to modification of the X-ray fluorescence analysis: the applicacation of X-ray radiometric analysis. In this case the soil sample is excited by radiation of radioisotope source. We used ^{241}Am as a source; as due to high energy of radiated line, it is possible to increase the energy range of fluorescence lines. In this regard, the mass of soil sample must be increased from 3-4 to appr. 8 g. Advantages of ^{241}Am isotope source compared to the X-ray tube are: high-energy radiation, hardly available by X-ray tubes; highly monochromatic radiation: high ratio of the emission line intensity to the continuum; high radiation stability (long half-life) and small geometric dimensions. Disadvantages of isotopic sources are follows: low intensity of radiation: standard intensity of the source is equal to $3.7 \cdot 10^{10} s^{-1}$; it is not possible to "switch off" the radiation source as the X-ray tube, it can only be shielded; problem of disposal of spent sources.

Thus, the content of La and Ce were determined in soils; however it required to increase the time savings of spectrum up to 15-20 minutes (Savichev and Vodyanitskii, 2009).

Analytical lines of the heavy lanthanides (Pr, Nd, Sm) are overlapped by β-components of the light lanthanides and barium. Soil Clarkes are not high for this group of elements (mg/kg): 7.6 (Pr), 19 (Nd), 4.8 (Sm). In order to determine Pr and Sm, when their content in soil are at least at the Clarke level, the spectrum set time has been increased up to 1.5-2 hours.

To determine the heavy lanthanides (Eu, Gd, Tb, Dy), we need to solve the problem of lines overlapping. We developed a method of X-ray radiometric analysis of the heavy lanthanides of yttrium group (Eu, Gd, Tb, Dy) in soils at concentrations above Clarke.

1.4. The Technogenic Portion of Metal

In this regard, it is important to know the technogenic portion of metal. Geochemists widely used the index of metal radial differentiation R compared to parent rock (Perel'man and Kasimov, 1999):

$$R = C_A : C_C,$$

where C_A and C_C are metal contents in A- and C-horizons. This indicator determines metal technogenensity, in case of aerial soil pollution.

However, this index is correct when soil profile is initially homogeneous, that is rarely observed. More often the the profile is lithologically heterogeneous, that is reflected in variation of granulometric distribution and content of heavy metals, associated with clay fraction. That's why for many soil types the indicator should be based on accounting of content of conservative minerals or elements-"witnesses." Quartz, zircon, tourmaline, and garnet can be used as minerals-"witnesses" to normalize conservative minerals.

Al, Ti or Zr can be used as conservative normalizing chemical elements. This is a basis for calculation of the adjusted coefficient of heavy metals enrichment (HME), when the content of HM is normalized to the content of Al (as shown bellow) as a conservative element, located mainly in the composition of aluminosilicate (Baron et al. 2006):

$$HME = (Me_A : Al_A) : (Me_C : Al_C),$$

where Me_A and Me_C are total contents of heavy metal (metalloid) in A- and C-horizons; Al_A and Al_C are total aluminum contents in A- and C-horizons.

This expression can be converted to calculate the ratio of metal technogenensity *Tg* as a percentage of its total content (Baron at al, 2006):

$$Tg = 100\% \cdot (HME - 1) : HME.$$

Contribution of natural biogenic processes into accumulation of most elements in the surface horizons does not exceed 20% of their total content. Therefore, *Tg* value > 20% was accepted as a limit value, that separates technogenic metal from natural one, accumulated biogenically. The profile analysis of elements' distribution allows somehow to solve the problem of heavy metals technogenensity without using background data.

2. Aerial Pollution of Soils by Micropollutants

2.1. Moscow City

2.1.1. The Lanthanides in Soils

In Moscow, the lanthanides content in urban soils rarely exceed the background values, excluding areas near highways. The degree of soil pollution depends on the intensity of traffic.

For example, in the West of Moscow (Minsk highway) soils are contaminated by the lanthanides less than in the East (Shchelkovo highway). Soil contamination riches the maximum at a distance up to 10 m from the highway. Total La-content in the soil at a distance of 5 and 10 m from Minsk highway was equal to 10.1 and 8.2 mg/kg, exceeding the background (6.85 mg/kg) by 1.5 and 1.2 times. Near Shchelkovo highway, La-content exceeded the background by 3.0 and 2.5 times. Ce-content in the soil at a distance of 5 and 10 m from Minsk highway was equal to 26.6 and 15.9 mg/kg, exceeding the background (14.5 mg/kg) by 1.8 and 1.1 times. Near Schelkovo highway Ce-content was exceeded the background by 2.6 and 2.3 times. The Sm-content in soil at a distance of 5 and 10 m from Minsk highway was equal to 3.98 and 2.12 mg/kg, exceeding the background (1.61 mg/kg) by 2.5 and 1.3 times. Near Schelkovo highway Sm-content exceeded the background by 2.8 and 2.7 times. Tb-content in the soil at a distance of 5 and 10 m from the Minsk highway was equal to 0.21 and 0.14 mg/kg, exceeding the background (0.14 mg/kg) by 1.5 times. Near Schelkovo highway Tb-content exceeded the background by 2.8 and 2.3 times.

2.1.2. PGM in Soils

Total PGM-contents in samples of soil and spatially related street dust are given in Table 1.

Determined elements can be subdivided into two groups. The first group consist of Ir and Ru, their content in soil and street dust is extremely low and random. The second group consists of Rh, Pd and Pt - the main components of catalytic exhaust neutralizers. Their content in most analyzed soils and street dust samples exceed Clarke level by many times; they have a strong variation and asymmetric frequency distribution.

On average, Rh-, Pd- and Pt-content in dust exceed their content in soil twice, indicating higher enrichment in street dust by technogenic compounds compared with soil samples (Ladonin, 2016). There is a close statistical relationship between the content of Rh, Pd and Pt in the soil and their content in street dust. The coefficients of determination R^2 of these elements are 0.58; 0.82 and 0.81 correspondingly. This may indicate that the input in soils of these elements occurs predominantly from a single source - combustion engine vehicles. The ratio of technogenic compounds of PGM is so great, that a natural variation in their content does not almost effect on the observed relationship. For Ru and Ir such a connection is not found.

Table 1. Total contents of PGM in soils and road dust of the South-Eastern administrative district of Moscow, ppb

Element	**Ru**		**Rh**		**Pd**		**Ir**		**Pt**	
	1*	**2****	**1**	**2**	**1**	**2**	**1**	**2**	**1**	**2**
Average	1.6	1.6	8.8	17.9	41.8	70.8	9.6	3.9	91.7	158.0
Median	1.6	1.6	6.7	13.3	34.2	51.7	9.5	4.3	88.5	132.6
Minimum	0.7	0.7	0.5	2.7	3.4	7.7	3.9	1.6	6.3	12.4
Maximum	2.4	2.2	29.3	54.5	112.2	225.3	13.9	6.2	183.9	356.6
Standard deviation	0.6	0.4	7.1	14.2	28.7	52.7	2.5	1.3	51.6	106.8

*1 - soils, ** 2 – road dust

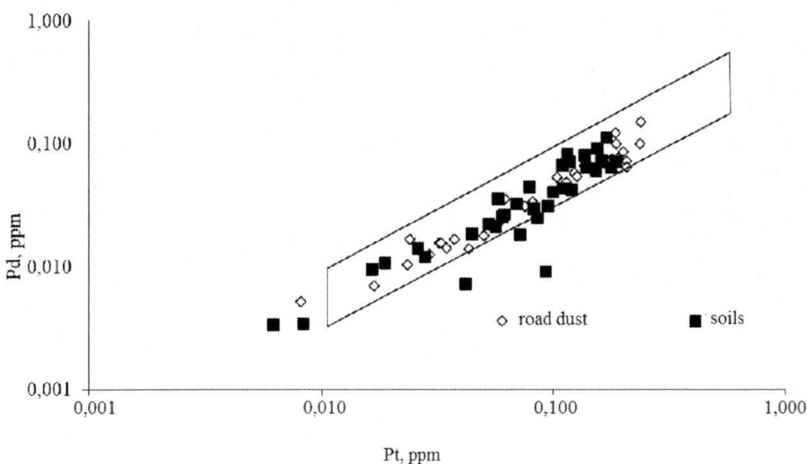

Figure 1. Pt/Pd-ratio in Moscow soils and road dust. The figure highlighted represents the ratios of the emissions of vehicles with catalytic exhaust converters.

Figure 1 is a graph of relationships between two PGM: Pt and Pd for soils and street dust samples. Highlighted area represents the ratio of emissions of vehicles with catalytic exhaust neutralizers (Ladonin, 2016). Relationships of PGM obtained for samples of soil and street dust are very close. Soils are contaminated by PGM from a single source, this pollution is modern and on-going. Thus the only source of PGM in Moscow city is combustion engine vehicles. Indeed, most of the data for soils are in the area, corresponding to Pt:Pd-ratio, typical for emission of combustion engine with catalytic exhaust neutralizers.

2.2. Cherepovets and Suburban Area

Key polluting elements from the Cherepovets steel mill "Severstal" are heavy metals: Cr, Ni, Cu, Zn (Doncheva et al. 1992). One source of the lanthanides is coal ash emitted from the local thermal power plant. Coal of Pechora basin, used at Cherepovets power plant, contains of 24 mg/kg Ce and 15 mg/kg La (Ivanov, 1974); at ash content of 20%, ash contains 120 mg/kg Ce and 75 mg/kg La, that is significantly exceeded their content in soil. In addition, the steel mill “Severstal” has some brunches, emitting lanthanides: sinter plant and blast furnaces.

2.2.1. Total Content of the Light Lanthanides (1993)

The first information on the lanthanides was obtained by XRF-method. We analyzed 16 samples of sod-carbonate soils, collected in four soil profiles located at different distances to the north of the steel mill “Severstal”: profile 1-2 km, profile 2-5 km, profile 3-8 km, profile 4-25 km. Four of the light lanthanides have been determined: La, Ce, Pr, Nd.

Table 2. Lanthanides contents (ppm) in sod-carbonate soils of the Cherepovets technogeochemical anomaly

Horizon	Depth	La	Ce	Ce:La	Pr	Nd	Sm
2 km from plant. Profile 1							
Ap	0-10	47	69	1.5	13	36	9
Ap	10-20	44	68	1.5	12	33	9
B1	30-47	50	74	1.5	14	38	10
Cca	70-85	29	43	1.5	8	20	-
Average	-	42	63	1.5	12	32	-
5 km from plant. Profile 2							
Ap	10-20	39	58	1.5	11	27	7
B1g	30-38	44	65	1.5	12	31	7
B2gca	38-60	35	46	1.3	9	24	6
Cgca	60-96	32	48	1.5	9	23	-
Average	-	37	54	1.4	10	26	-
8 km from plant. Profile 3							
Ap	0-10	44	62	1.4	12	32	8
Ap	10-20	39	64	1.6	10	27	7
AB	30-42	49	66	1.3	13	34	8
BCca	66-85	36	57	1.6	10	25	6
Average	-	42	62	1.5	11	29	7
25 km from plant. Profile 4							
Ap	0-10	42	55	1.3	11	28	7
Ap	10-20	40	57	1.4	10	26	7
B1	30-51	35	51	1.5	9	25	6
BC	53-66	36	56	1.5	10	26	7
Average	-	38	55	1.4	10	26	7
Clarke	-	26	52.2	2.0	6.0	22	4.3

Table 3. Percentage ratio of technogenensity *Tg* of lanthanides in soils of the Cherepovets technogeochemical anomaly (% from total content)

Profile, distance from the plant	La	Ce	Pr	Nd
1. 2 km	**43***	**42**	**39**	**41**
2. 5 km	**23**	**22**	**21**	**24**
3. 8 km	**25**	17	16	18
4. 25 km	18	3	7	5

*highlighted is the significant technogenensity with $Tg > 20\%$

The average content of lanthanum and cerium in soils is above Clarke (Table 2). As soil samples were collected within the area of technogeochemical anomaly, it is important to divide the total metal content into natural and technogenic parts. Two geochemical indexes: Tg and Tg^{Ca} were used.

Application of index of technogenensity T_g lead to underestimation of the proportion of metals technogenensity in profile 2, located relatively close to the impact source: up to T_g = 0-3%. Possible reason is that the soil was formed on silicate-carbonate moraine: the total CaO-content in Cgca-horizon is 10.2%. Application of adjusted technogenensity index Tg^{Ca} in profile 2 raises the percentage of technogenic metals up to 17-23%, which corresponds with profile location within geochemical catena (Table 3).

Percentage ratio of technogenensity T_g^{Ca} for lanthanum decreases in the following order: 43 → 23 → 25 → 18% with the distance from the plant. Thus, the topsoil near "Severstal" steel mill is polluted by lanthanum in a greater extent.

Percentage ratio of technogenensity T_g^{Ca} for cerium reduced as follows: 42 → 22 → 17 → 3% with the distance from the plant. Maximum extent of cerium contamination is determined in the topsoil near the steel mill.

For neodymium and praseodymium the technogenensity T_g^{Ca} reducing with the distance from the steel mill has qualitatively the same tendency as for lanthanum and cerium.

The general distribution of the technogenensity percentage ratio for rare heavy metals was as expected. Significant part of the lanthanides in soils near the steel mill has anthropogenic origin.

2.2.2. The Content of All Lanthanides (2013)

Later, all 14 lanthanides have been determined using ICP-MS method (Ladonin, 2016). Samples were gathered in five profiles of industrisols (U-horizon) and sod-carbonate leached loam soils (horizons A, B and C, located to the north from the steel mill). Profile 1 (industrisol) is directly adjacent to the territory of the "Severstal" steel mill. Profile 2 is located at a distance of 2 km from the steel mill, profile 3-6 km, profile 4-12 km and profile 5-30 km. Soils are sod-carbonate leached. Humus-accumulative horizon has sufficiently high content of humus, slightly acidic reaction turns into slightly alkaline down the profile. Near the source of pollution slightly alkaline reaction is registrated throughout the profile, because of the input of anthropogenic compounds.

To neutralize the differences, connected with elements' abundance in the Earth's crust, the results of total content determination of the lanthanides were normalized on clay of the Russian Plain (Migdisov et al. 1994). Normalized total content of the lanthanides are shown on Figure 2.

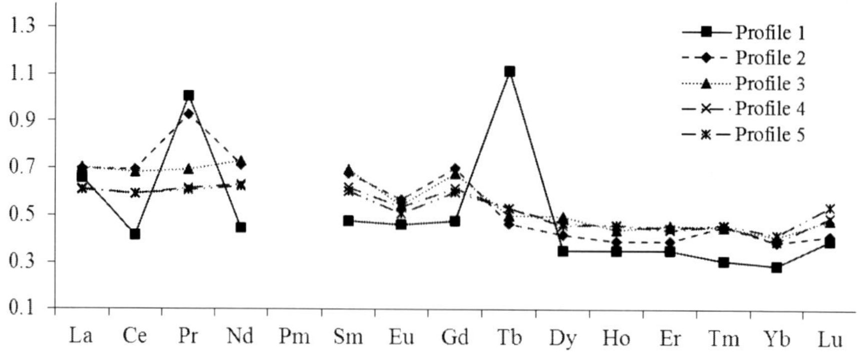

Figure 2. Total lanthanides content in upper (0-10 cm) layers of the Cherepovets technogeochemical anomaly soils (normalized on the clay of Russian Plattform).

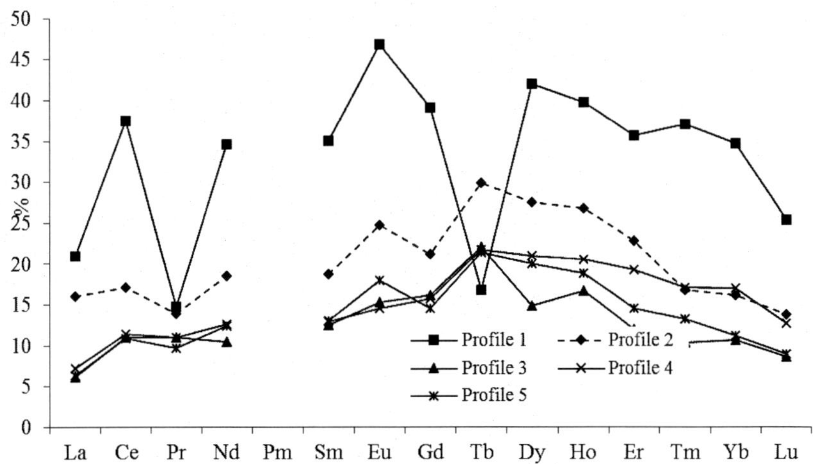

Figure 3. Percentage ratio of nitric acid-soluble lanthanides in soils of the Cherepovets technogeochemical anomaly (% of the total content).

The technogenic pollution of soils modifies the content of the lanthanides significantly. First of all - the increase of praseodymium and terbium content. Maximum increasing of terbium is observed in the profile 1. Increasing of praseodymium content in this profile is less, but in contrast to terbium, is registrated also in the profile 2. This indicates significant changes in element composition of atmospheric fall-outs with increasing of a distance from the source of pollution. Similar to praseodymium, in the soil of profiles 2, there is a tendency of increasing of the content of other light lanthanides: from lanthanum to gadolinium inclusive.

Then we determine lanthanides' compounds in the same soil collection.

2.2.3. Acid-Soluble Lanthanides Compounds

The degree of the lanthanides dissolution by 1 M nitric acid is shown on Figure 3. There is a significant increase in the extraction rate of all these elements, except praseodymium and terbium in the most contaminated soil profile 1 and a less increase in the degree of elements extraction from lanthanum to erbium in the profile 2. There is also a tendency to increase the degree of extraction from soil for elements of the middle of the lanthanides group: from neodymium to erbium. This trend is connected with combination of two factors: the higher affinity of the heavy lanthanides to soil components (primarily, due to adsorption), that reduces the dissolution of the heavy lanthanide by nitric acid, and a decrease in soil technogenic pollution from the light lanthanides to the heavy lanthanides.

The lanthanides can be subdivided into three groups according to dissolution by nitric acid. The first group includes Pr and Tb, which are minimally soluble by acid in soils near the steel mill, where their total content has significantly increased. The second group includes mainly light lanthanides: La, Ce, Nd, Sm, Eu, Gd, Dy, Ho and Er; their extraction by acid is high in soils near the steel mill. The third group includes the heavy lanthanides: Tm, Yb and Lu; their solubility is weak and is observed only in the most polluted profile 1.

Based on the grouping of elements, it is concluded that praseodymium and terbium of technogenic origin, found in soils near pollution source, are presented in technogenic particles of large size, hardly soluble in the nitric acid, and falling quickly onto the soil surface. The lanthanides composition of smaller and lighter particles of technogenic origin, spreading far from the source of pollution, is more equable, and elements of these particles are readily extracted by nitric acid. As the element number rises, the extent of its involvement into technogenic substances flows is gradually reduced.

2.2.4. Fractional Chemical Composition of the Lanthanides Compounds

Residual fraction (80-95% of the total content) prevails significantly over the other fractions for all the lanthanides in all soils, regardless of the distance from the source of pollution. This means, that the main soil components, determining background level of the lanthanides content in the soil, are aluminosilicate minerals, in which structure the lanthanides are strongly fixed, and the lanthanides-containing technogenic compounds are sufficiently chemically stable.

Influence of soil factors on the redistribution of the lanthanides between other soil constituents is insignificant. Organic matter (Figure 4) effects most noticeably on the lanthanides fractional composition. The largest percentage of the fraction, associated with organic matter, is in the middle of the lanthanides group. Both light and heavy lanthanides have less affinity to an organic substance.

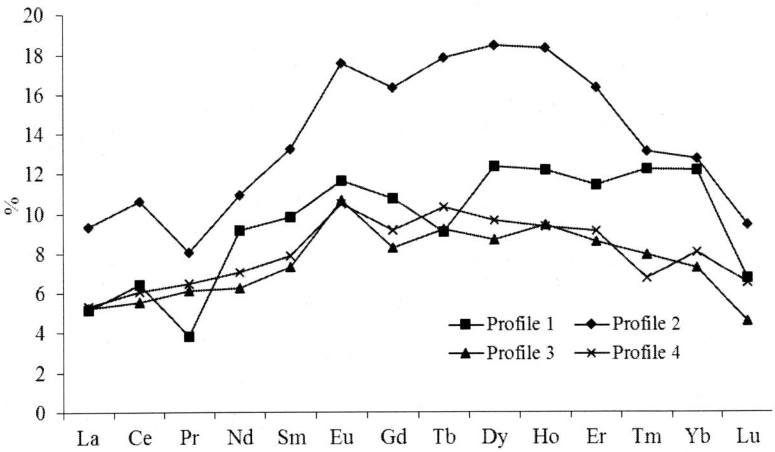

Figure 4. Fraction of lanthanides, associated with organic matter (% from total content) in soils of the Cherepovets technogeochemical anomaly.

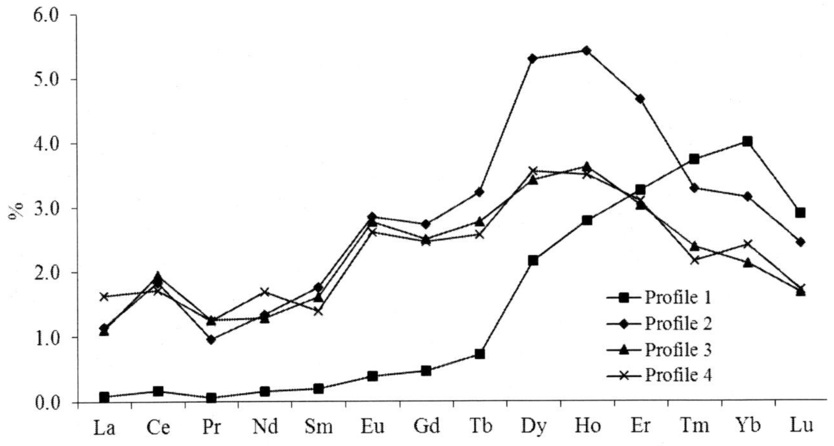

Figure 5. Fraction of lanthanides, associated with Fe and Mn (hydr)oxides (% from total content) in soils of the Cherepovets technogeochemical anomaly.

The content and the percentage of fraction, bound to (hydr)oxides of Fe and Mn, is significantly smaller than fraction, associated with organic matter, ranging from 0.1 to 5% of the total content. In contrast to the fraction, associated with organic matter, the maximum increase in the percentage of fraction, bound to (hydr)oxides of Fe and Mn is shifted in the contaminated soils profiles towards heavy elements with a maximum of ytterbium, dysprosium and holmium. This indicates, that the affinity of the lanthanides to iron oxides, which are the main components of emission substances, increases with the serial number of the lanthanide. In less polluted and background soils, there is more noticeable decline in the fraction percentage ratio of the heavy lanthanides. This may be due to the fact, that the content of technogenic heavy lanthanides are not significant. In addition, it is possible the increase of fixing of the lanthanides on ferrous minerals surface with the increase of their serial number, and the extraction capability by Tamm's solution may be insufficient for their complete extraction.

2.2.5. Platinum Group Metals in the Soil of the Steel Mill Affected Zone

Total content of PGM in soils, located far from the steel mill, are close to Clarke. Near the pollution source, the total content and the content of acid-soluble forms of PGM increase significantly (Table 4).

Points on Figure 5, are reflecting the ratio Pt to Pd in soils near the steel mill, are located away from areas of PGM-ratios, obtained for Moscow city road dust. This means a little current contribution of vehicle emissions to soil pollution by PGM in the area influenced by the Cherepovets steel mill.

2.3. Revda

Revda technogeochemical anomaly is located near Pervouralsk-Revda industrial zone in the Sverdlovsk region. The plant is located in the area of Pervouralsk-Revda industrial hub in the Sverdlovsk region; it works since 1940. There are currently two main workshops: copper and sulfuric acid. Atmospheric plant emissions contain sulfur dioxide, fluoride hydrogen, and aerosols include a number of heavy metals: Cu, Zn, As, Cd, etc. (Vorobeichik et al. 1994). On the “Khomutovka” territory, where the study was conducted, soils are heavy clay gray forest.

Area gradation on the reaction to technogenic impacts was made according to vegetation state. In the area of technogenic desert trees were completely killed, herbal layer was either absent or it consists of horsetail and cereals, strongly moss cover was developed. Litter and humus soil horizon were completely washed away on eroded areas. Leaves from trees were burned and tops were dried in impact zone. Buffer zone is divided into two: the near and far to the plant, according to the vegetation state. Vegetation is weakly and moderately depressed, there is dieback of coniferous trees in the near buffer zone. Vegetation is not damaged in far buffer zone (Vorobeichik et al. 1994).

We analyzed the soil in four profiles in accordance with the ecological gradation of area. Samples were selected in 2000. Profile 1 was opened in a technogenic desert at 0.5 km to the east of the plant to the prevailing wind rose; profile 2 - in the impact zone, 1 km to the west of the plant; profile 3 - in the near the buffer zone of 7 kilometers to the west of the plant; profile 4 - in the far buffer zone of 30 km to the west of the plant (Vodyanitskii et al. 2010b). Soil is strongly acidified (pH_{H2O} = 4.6) in the upper layer in the technogenic desert (profile 1). Soil acidity decreases according to the distance from the plant.

We emphasize, that the pure lands in the Sverdlovsk region and in other regions of Urals are practically absent; soils are contaminated everywhere: the influence zone of one plant overlaps often the influence zone of another one. Also this applies to Revda technogenic anomaly. We used as background the published data on the content of heavy metals in the A1-horizon (Vishera reserve in the north of the Perm region, on the border with the Sverdlovsk region) (Voronchihina and Larionova, 2002).

The main polluting elements of copper smelter are: Cu, Zn, As (Vodyanitskii et al. 2011). Total the lanthanides content in soils were studied by XRF-method.

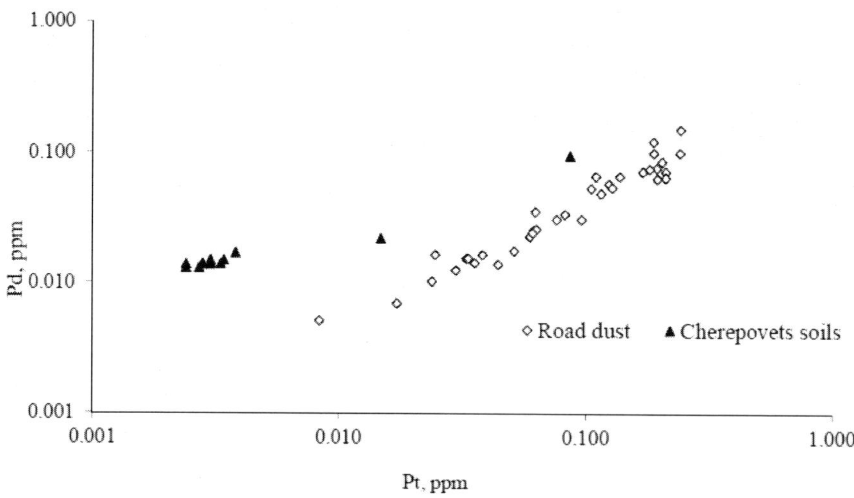

Figure 6. Pt/Pd-ratio in soils of the Cherepovets technogeochemical anomaly and Moscow road dust.

The average Ce-content in soils is considerably higher Clarke (Table 5). Since the soil samples are taken in the territory of technogeochemical anomaly, it is important to establish, to what extent the content of the lanthanides is genetic (natural-geochemical) basis, and to what extent - technogenic. The T_g-index was used for this purpose.

The average Y-content is higher in buffer zone (35-39 vs. 27-34 mg/kg near the plant) that can be attributed to the territory heterogeneity of the Revda geochemical anomaly. At a distance of 7-30 km from the plant, bedrocks differ from surrounding bedrocks, there are increased content of at least two the lanthanides (Y and Ce). This is especially noticeable for cerium. Its average content is much higher in the buffer and background zones: 82-83 against 50-61 mg/kg near the plant. In this regard, we call attention to the high Ce:Y-ratio in soils of the Revda geochemical anomaly, which varies from 1.5 to 2.4, exceeding the Clarke module (2.3).

Table 4. PGM contents in soils of the Cherepovets technogeochemical anomaly (ppb)

Profile	Horizon	Ru		Rh		Pd		Ir		Pt	
		1*	2**	1	2	1	2	1	2	1	2
1	U	4.3	0.114	48.9	0.551	95	11.52	12	0.10	85.6	6.25
2	A	≤0.2	≤0.003	4.6	0.006	22	3.43	6	0.08	14.7	0.65
	B	≤0.2	≤0.003	0.4	0.008	17	3.67	3	0.07	3.8	0.06
	C	≤0.2	≤0.003	0.3	0.006	15	3.78	3	0.07	3.0	0.04
3	A	≤0.2	≤0.003	0.4	0.008	15	3.15	3	0.07	3.4	0.07
	B	≤0.2	≤0.003	0.3	0.007	14	3.43	3	0.07	3.3	0.05
	C	≤0.2	≤0.003	0.3	0.006	14	3.25	3	0.07	2.8	0.04
4	A	≤0.2	≤0.003	0.3	0.007	14	3.06	3	0.07	3.0	0.05
	B	≤0.2	≤0.003	0.3	0.006	13	3.15	3	0.07	2.7	0.04
	C	≤0.2	≤0.003	0.2	0.006	13	3.33	3	0.07	2.4	0.04
5	A	≤0.2	≤0.003	0.3	0.007	14	3.20	3	0.07	3.0	0.05
	B	≤0.2	≤0.003	0.3	0.007	14	3.38	3	0.07	2.8	0.04
	C	≤0.2	≤0.003	0.3	0.006	14	3.44	3	0.07	2.4	0.04

* 1 – total content, **2 – 1 n. nitric acid soluble forms

Table 5. Lanthanides contents in soils of the Revda technogeochemical anomaly (ppm)

Horizon	Depth	Y	La	Ce	La/Y	Ce/La	Ce/Y	Pr	Nd	Sm
1. Technogenic desert (0.5 km from the plant)										
A0A1	0-2	29	37	54	1.3	1.5	1.9	11	28	7
AB1	2-13	26	39	67	1.5	1.7	2.6	12	29	7
B	13-40	26	34	63	1.3	1.8	2.4	10	26	6
2. Impact zone (1 km from the plant)										
Average	-	27	37	61	1.4	1.7	2.3	11	28	7
A0	0-2	24	29	38	1.2	1.3	1.6	9	20	5
A1g	2-10	36	35	53	1.0	1.5	1.5	11	26	6
A2g	10-22	27	32	59	1.2	1.8	2.2	11	25	6
B1g	22-44	29	34	50	1.2	1.5	1.7	10	26	6
B2	44-63	40	26	46	0.6	1.8	1.1	8	19	-
BC	63-70	48	33	54	0.7	1.6	1.1	10	25	6
Average	-	34	31	50	0.9	1.6	1.5	10	24	6
3. Buffer zone (7 km from the plant)										
A1/	6-11	39	36	65	0.9	1.8	1.7	10	27	7
A1//	11-18	39	48	86	1.2	1.8	2.2	14	38	9
A2B	18-32	37	43	89	1.2	2.1	2.4	13	33	7
B1	32-50	38	43	85	1.1	2.0	2.2	14	35	8
B2	50-63	41	42	83	1.0	2.0	4.0	13	34	8
BC	63-70	42	48	89	1.1	1.8	2.1	14	37	7
Average		39	43	83	1.1	1.9	2.4	13	34	8
Clarke*		23	26	52.2	1.13	0.50	2.27	6.0	22	4.3

* Kabata-Pendias (2011)

Table 6. Percentage ratio of technogenensity *Tg* of lanthanides in soils of the Revda technogeochemical anomaly(% from total content)

Profile, zone	Y	La	Ce	Pr	Nd	Sm
1. Technogenic desert	**26**	**25**	6	**29**	**24**	**23**
2. Buffer zone	**30**	**51**	**47**	**49**	**48**	**51**
3. Buffer zone	**32**	17	15	16	18	17

*highlighted is the significant technogenensity with $Tg > 20\%$

When calculating the technogenensity index T_g, attention was drawn to the fact, that a maximum of rare earth metals falls on the humus horizon A1g in impact zone. According to this fact, we counted the value of technogenensity index T_g.

As can be seen from Table 6, Ce is the low-technogenic metal, technogenensities of lanthanum and the other the lanthanides are higher. In the buffer zone, the technogenensities of all studied the lanthanides are unreliable.

The highest proportion of the lanthanides technogenensity is not in technogenic desert zone (profile 1), but in the impact zone (profile 2). It is generally observed for other heavy metals and metalloid. Indeed, in the litter (horizon A0) in the technogenic desert and in the impact zone, accumulated arsenic is respectively equal to: 257 and 1106, lead: 1148 and 4564, copper: 1476 and 8730, zinc: 969 and 2529 mg/kg.

The content of heavy elements in the impact zone is 2-6 times higher than in the technogenical desert. This difference affects technogenensity of heavy elements. Technogenensities of As, Pb, Cu, Zn amounted to 76-85% in the technogenic desert, whereas in the impact zone they reached 99-100%. What is the reason for the relatively weak accumulation of heavy elements in the technogenic desert? Perhaps, technogenic desert was formed on the slope, from which the surface layer of contaminated soil was washed away by water erosion. Here, at a depth of 2 cm, begins illuvial horizon. But there are other reasons, for example, the transfer of aerosols through the desert.

Thus, in the territory of Revda anomaly, technogenensities of Y, La, Ce are lower than ones of As, Pb, Cu, Zn, but comparable with technogenensity of Ga (90%) and higher than one of Ni (50%).

3. Hydrogenic Pollution of Alluvial Soils by Micropollutants in Perm

3.1. Objects

Hydrogenic soil contamination is studied much worse than aerial one, because of the smaller scale. Meanwhile more local hydrogenic pollution can be quite dangerous. Raw sewage, falling into small rivers, pollutes the limited soil mass, but with high concentrations of pollutants, whereas emissions are dispersed over a large area at aerial pollution. Alluvial soils are enriched in sludge, contaminated by heavy metals and metalloids (Osovetskii and Men'shikova, 2006). In the Ural region, the main technogenic pollutants of river sediments are Bi, Ag, Zn, Pb, Cu, W, Ni (Osovetskii and Men'shikova, 2006). As a result of untreated sewage discharge, contamination of alluvial soils in the floodplain of the rivers, especially small with a low capacity for dilution, may be greater than in soils of urban areas on automorphic positions.

The fate of heavy metals largely depends on soil neoformations (Fe-Mn nodules): elliptical ortsteins and tubular röhrensteins, that have the ability to remain the metals from biological cycle by sorption and lasting fixing them by (hydr)oxides of iron and manganese. Binding metals prevents them from getting into the soil solution and then biologically in organisms and further into the biological cycle.

Let's consider the pollution of alluvial soils from untreated industrial effluent, entering the small rivers, tributaries of Kama river. Industrial waste from Perm city contains high concentrations of Cu, Pb, Zn, As, Mo, Ni, Cd, Hg (Schukova, 2005). Studies (involving M. N. Vlasov and A. V. Kozheva) were carried out on alluvial soils of the small rivers of Perm, which are flooded during the flood and are covered by sludge. Profiles are placed in the zone of influence of Perm-Krasnokamsk industrial hub, in left tributaries of the Kama river: Iva, Yegoshikha, Danilikha, Mulyanka and right one (Las'va).

Profiles of grey humus gleyed typical soils were opened in the floodplains of Iva, Yegishikha, Danilikha and Mulyanka rivers. Profile of grey humus gleyed soil was opened in the floodplains of Las'va river (1 km downstream from Krasnokamsk town). The studies were conducted in 2006; 5 profiles were opened; 20 soil samples and 10 samples of nodules were analyzed (Vodyanitskii et al. 2008).

All studied soils (but the soil of Iva valley) are subjected to anthropogenic pressure. The worst water quality was observed for the lower reaches of Yegoshikha and Danilikha rivers, the water is not drinkable. According to a report of the Environmental Protection Agency in 2004, the average water quality in the Iva river corresponds to Class 2 (net), in the Yegoshikha estuary - 3 (moderately polluted), in the Danilikha river - 6 (very dirty), in the Mulyanka river - 2-3, and in the upper reaches under the high content of nitrates and iron - 4 (Status and Protection ..., 2004).

3.2. Light Lanthanides in Alluvial Soils of the City

Total the lanthanides content in soils were studied by XRF-method. Alluvial meadow-bog soils were studied in the city of Perm in the floodplains of small rivers, as well as Kama-river. We studied both fine earth and iron-manganese nodules, separated from it: ortsteins and röhrensteins. The results are shown in Table 7.

Let us first consider the content of lanthanum and cerium in floodplain soils of clean rivers: Obva and Kama. Fe and Fe-Mn new substances: röhrensteins and ortsteins are formed in them. La-content in fine earth of these soils varies slightly from 28 to 41 mg/kg, which is slightly greater than the European Clarke (26 mg/kg). La-content in röhrensteins is even lower: 14-25 mg/kg. But La-content in ortsteins in gley horizon (31-55 cm) increased up to 56 mg/kg. As a result, the La-concentration coefficient in the nodules varies considerably: from 0.3-0.7 in röhrensteins to 1.5 in ortsteins.

The Ce-content in the fine earth of unpolluted soil varies from 41 to 60 mg/kg, which is close to the European Clarke (52.2 mg/kg). Ce-contents in röhrensteins are very contrasting: from 16-38 mg/kg in the soil of the Obva floodplain to 191 mg/kg in the soil of the Kama river floodplain. The cerium accumulation and a high Ce:La-ratio (3.4; Clarke Ce:La-ratio = 2.0) indicates a possible contamination of Kama water by cerium in the profile 41.

Table 7. La and Ce contents in the fine earth and nodules in the clean and polluted alluvial soils of the Urals (ppm)

Horizon, depth, cm	Matherial	La	K_{conc} (La)	Ce	K_{conc} (Ce)	Ce/La
Humus-gley typical soil in the floodplain of the clean Obva river, profile 51						
C2g, 37-75	fine earth	37		58		1.6
	röhrensteins	25	0.7	38	0.6	1.5
G~~, 75-90	fine earth	41		60		1.5
	röhrensteins	14	0.3	16	0.3	1.1
Layered typical soils in the floodplain of the clean Obva river, profile 53						
C2~~, 20-27	fine earth	28		41		1.5
C6~~, 71-78	fine earth	26		38		1.5
Humus-gley mineralized soils in the floodplain of the polluted Kama river, profile 41						
G~~, 31-55	fine earth	38		57		1.5
	ortsteins	56	1.5	191	3.3	3.4
Gleyed agrosol in the floodplain of the polluted Mulyanka river, profile 33						
C2~~, 49-75	fine earth	31		47		1.5
	ortsteins	104	3.3	324	6.9	3.1
C3~~, 75-107	fine earth	31		46		1.5
	ortsteins	108	3.5	302	6.6	2.8
C4g,t~~, 107-137	fine earth	34		45		1.3
	ortsteins	100	2.9	243	5.4	2.4
C5g~~, >137	fine earth	30		48		1.6
	ortsteins	86	2.9	150	3.1	1.7
Average			2.2		3.7	
Clarke		26		52.2		2.0

The different situation is in gley agrosol in the floodplain of contaminated Mulyanka river. Oil refinering company is one among enterprises, discharging poorly treated wastewater. Fe-Mn-ortsteins are formed in the contaminated floodplain soils of the Mulyanka river.

In glay agrosol in the floodplain of contaminated Mulyanka river, Ce-content in the fine earth varies slightly: from 38 to 60 mg/kg. In contrast, in the composition of nodules Ce-content varies greatly: from 16 to 324 mg/kg. As a result, the Ce-concentration coefficient in the nodules is ranged from 0.3 to 6.9.

Let's see how the Ce-content is distributed along the gley alluvial agrosol profile in the floodplain of Mulyanka river. The Ce-content in the fine earth on the profile remains practically unchanged: 45-48 mg/kg (below Clarke - 52.2 mg/kg). But the Ce-content in ortsteins varies greatly from 150 mg/kg in the most low horizon C5g ~~ to 324 mg/kg in the top of studied horizons C2 at a depth of 49-75 cm. Obviously, cerium as the most active lanthanide is accumulated in ortsteins in the top of the soil column.

Different formation mechanisms of röhrensteins and ortsteins affect the values of the lanthanides concentration coefficients. Röhrensteins, formed by the participation of organic root exudates, in the floodplain of the clean Obva river are depleted by lanthanum and cerium (K_{conc} = 0.3-0.7). Probably, organic ligands in röhrensteins are spent on fixing iron, which dominates in the röhrensteins composition; in particular, this is reflected in the extremely high values of Fe:Mn-ratio, it reaches 25-100 in röhrensteins. On the other hand, Fe-Mn-ortsteins, formed by alternating redox regime, are enriched in lanthanides. For the Fe-Mn-ortsteins, there is a uniform accumulation of elements, particularly a low ratio of Fe:Mn-ratio (1.4-12).

The degree of river pollution is very important. The soil in the floodplain of the Mulyanka river are heavily polluted. Accumulation coefficient K_{conc} of lanthanum in the floodplain concretions reaches 2.9-3.5 and accumulation coefficient K_{conc} of cerium - 3.1-6.9. Dirty water of small river Mulyanka, coming in abundant Kama, is diluted. As a result, ortsteins in the floodplain soil of Kama river are less contaminated. Lanthanum accumulation coefficient K_{conc} is equal to 1.5, and for cerium K_{conc} = 3.3 in these nodules.

Maximum cerium accumulation in nodules is not by chance, cerium is very sensitive to changes in the redox regime. Weak lanthanum accumulation is due to its relative physical and chemical inertness.

Let's consider Ce:La-ratio in the fine earth and nodules. This ratio is practically constant in fine earth (average is 1.5), which is close to Clarke ratio (1.9). In nodules it varies from 1.1 to 3.4, averaging 2.3. The fact, that the average Ce:La-ratio in the nodules is higher than in the fine earth, means that the cerium accumulation in nodules is more active than lanthanum one.

3.3. Heavy Lanthanides in the City Alluvial Soils

In the nodules, formed in the alluvial soils in the floodplains of the rivers Kama and Mulyanka, europium and terbium content are below the detection limit. The content of gadolinium and dysprosium are above the detection limit. Content of gadolinium is 13-16 mg/kg, and dysprosium - 11-13 mg/kg (Table 8). This is substantially more than Clarke values (6.1 and 4.5 mg/kg). Such high values of the lanthanides enrichment in nodules can be associated with anthropogenic contribution. The lanthanides are used as catalysts in petroleum refining. As a result, the lanthanides input rivers with sewage. In Perm, the local refineries, dumping into rivers poorly treated wastewater, pollute their by the lanthanides (such as Gd and Dy), which are fixed in floodplain soils in nodules composition.

A similar situation is observed in other rivers near the refineries. Refineries, located on the banks, are a major pollution cause of sediments in Rhine estuary (Sneller et al., 2000). For example, Ce-content exceeds 100 mg/kg in sediment.

Gd:Dy-ratio is close to Clarke in ortsteins. But Y:Gd- and Y:Dy-ratios are below Clarke. This is quite different from the situation in the Khibiny-Lovozero province, where these ratios were higher than Clarke (Vodyanitskii et al. 2010a). Lanthanides ratios show a fundamental difference in formation of these soil objects: soil formation on the rock with a high proportion of yttrium in the Khibiny-Lovozero province or the lanthanides binding in nodules in Cis-Urals.

The overall situation on the concentration of studied lanthanides in nodules of contaminated alluvial soils are presented in Table 8. If the average coefficients of concentration for La, Pr, Nd, Sm, Gd, Dy varies from 2.5 to 3.1; for the cerium it reaches a value of 5.5. The maximum accumulation of cerium as compared with the other the lanthanides is due to its sensitivity to a change in redox regime.

Table 8. Yttrium and heavy lanthanides contents in Fe-Mn nodules of the polluted alluvial soils of Urals (ppm)

Horizon	Depth, cm	Y	Gd	Dy	Gd:Dy	Y:Gd	Y:Dy
Profile 41. River Kama floodplain							
G~~	31-35	Not det.	15	11	1.36	Not det.	Not det.
Profile 33. River Mulyanka floodplain							
C2~~	49-75	32	16	12	1.33	2.0	2.7
C3~~	75-107	42	15	13	1.15	2.8	3.2
C4g~~	107-137	37	13	11	1.18	2.8	3.4
C5g~~	>137	40	14	11	1.27	2.8	3.6
Clarke		31	6.1	4.5	1.35	5.1	6.9

4. Contamination of Peat Soils by the Lanthanides from Spilled Oil at the Points of Its Production in Western Siberia

4.1. Objects and Methods

Oil and petroleum products are a powerful source of contamination. In the fields of oil production all components of the environment are changed: soil, biota, groundwater and surface water, the air. The environment components suffer in varying degrees. This is reflected in the structure of their protection costs. In the United States in the exploration and production of oil, costs for the protection of soil, air and water correspond to ratio 1: 5: 35 (Solntseva, 1998). A similar relation between costs takes place in Russia. The fact, that costs for protection of water resources exceed greatly the cost for the protection of soil, said about the special importance for the preservation of clear waters. Their pollution leads to the death of aquatic flora and fauna, as well as to diseases of people, drinking contaminated water. Under contamination of water, pollutants, extending away from the source, cause damage over a wide area: local pollution is transformed into a regional.

Currently marsh landscapes, occupying about 40% of the Middle Ob' area, are experienced ever-increasing anthropogenic pressure due to oil production, increasing the length of the communications, wearing pipelines on long-cultivated fields, many of which are just as most wetlands. Peat soils are influenced not only by petroleum hydrocarbons, but related mineral components: salts and heavy metals (Solntseva, 1998). Raised bogs, dominating in the region, are affected particularly.

There are three major sources of soil pollution in the fields of oil production: oil, saline formation water and drill cuttings stored in barns. We consider only the first two sources.

The oil contains metals such as V, Ni, Cr, Zn (Pikovskii, 1993). Furthermore, there are also other metals, as well as salts.

Formation of water, including the components of the drilling mud, contains salts, petroleum products, heavy metals; there are a high concentrations of alkali and alkaline earth metals and halogens (Sharova, 2009).

The degree of contamination depends not only on the content of pollutant (oil), but also on the type of soil. For example, in mineral soils with high background metals content, background excess by heavy metals is often negligible in areas of the oil spill (Fiedler et al. 2009). Peat soils are contaminated much stronger, particularly oligotrophic peats with low background content of heavy metals.

The field of oil production is located in the eastern part of the Middle Ob' Lowland with bogs, on the border of northern taiga and middle taiga subzone. In the studied area, raised (oligotrophic) ridge-hollow bogs occupy watershed flat surfaces.

Three studied sites have been transformed to varying degrees as a result of salt and oil pollution. Contaminated sites are located in an area within a radius of 16 km from the center of the field. The objects were selected on the basis of pollutants differences: as predominantly contaminated with oil and formation waters. Total projective cover (TPC) was determined at each site, the nature of violations of vegetation was analyzed.

Ash content of pure peat is only 1-2%. Meanwhile, the ash content in the contaminated soil reaches 47%, which increases peat pollution by chemical elements. The increase in the ash content of peat at pollution is due to the fact, that crude oil spreads, containing components of drilling mud, and its ash content is higher than the commercial oil. At the site of the oil spill, oil fractionates: light fractions evaporate and migrate, while the share of heavy fractions is increased, forming resinous-asphaltene crust with high ash content. The increase in the ash content, as a result of peat salinity by formation water, is not so much. In the course of the self-healing, ash content of peat soil decreases, approaching the background, though it does not reach.

The studies were conducted in July 2010. Peat samples were taken from test site by envelope method from depths 0-10, 10-30 cm (and in some places and up to 1 m). In the laboratory they were dried to air-dry state, and then incinerated in a muffle furnace.

The content of chemical elements in peat ash was determined by XRF-method on the “Respect” device, it also was examined the content of the lanthanides by X-ray radiometric method (Savichev and Vodyanitskii, 2009; Savichev and Vodyanitskii, 2011).

4.2. Laboratory Test for Distinguishing Contamination Nature

The nature of soil contamination is extremely heterogeneous: some are contaminated mainly by oil, the other are contaminated by the salty reservoir water. The dominance in soil of oil or minerals can be identified on the basis of chemical analysis of peat. An essential companion of oil is

Ni, and Br for saline solutions. We have proposed a simple laboratory test for distinguishing modern nature of peat contamination: the Ni:Br-ratio in peat ash. When the Ni:Br-ration is more than 1.5, peat is contaminated by oil preferably; when Ni:Br-ratio <0.7 - saline, and when 0.7 < Ni:Br < 1.4 - mixed contamination. We have considered only two contrasting type of peat contamination: by oil or by salty solutions.

But such a division of soils is not enough. When analyzing the chemical composition of contaminated peat ash, we drew attention to the strong variation in the content of many elements. This suggests heterogeneity of the samples, including the objects of different composition. To improve the homogeneity, each group of the samples (soil contaminated with oil or formation water) were divided into two groups, depending on the pollution ago. Self-healing of the soil is reflected in a change in its chemical composition. In terms of different pollutions, chemical composition of soils varies differently in time.

Iron is the sensor element for predominantly oil-contaminated soils. Soil, contaminated with oil recently, is strongly enriched in iron, but over time, as it happens in the background and peat bogs, Fe migrates actively, and oil-polluted peat loses it. For these conditions critical Fe_2O_3-value is equal to 8% in peat ash: in a recent (within 5 years) contaminated peat more than 8%, and for a long-standing pollution (over 10 years) it is less than 8%.

It is convenient to estimate the degree of soil salinity by formation waters, using the content of chlorine in the peat ash. The average Cl-content in the fresh salted peat is more than 1%, in desalted one under 1%. In the analysis of the chemical composition of peat ash, polluted by reservoir water, as well as contaminated oil, attention is drawn to a strong variation in the content of elements. This indicates a sample heterogeneity. We shared a generalized sample into two groups on the Cl-content. In the first group (n = 5), we collected samples, contaminated by oil recently (up to 5 years) and with high Cl (an average of 5.2%). The second group (n = 4) include the soils, contaminated for a long time (over 10 years) by formation water and desalted significantly. The Cl-content in these soil ash is low: an average of 0.2%.

Using the magnitude of the Student's *t*-test, we evaluated reliability of differences from the average content of each chemical elements in the two groups of soil from the background, and the differences between them.

4.3. The Lanthanides

Currently, geochemists include the lanthanides to the inactive complex-forming elements, partially migrating in strongly acidic and highly alkaline waters. They are allocated to the group of inactive elements, due to the small content in the soil and groundwater (Perel'man and Kasimov, 1999). There are several reasons for this: low Clarkes for these elements, as well as the poor solubility of minerals. However, the low mobility of the lanthanides are not always manifested. Lithogenic lanthanides phosphates are leached significantly from podzolic soils (Tyler, 2004), which calls into question the idea of their low migration capacity in the forest zone.

From our data (Table 9), soils, contaminated with oil, are greatly enriched in the lanthanides. In recently contaminated peat Y-content is twice the background, lanthanum is almost seven times, cerium is nine times.

It is noteworthy, that in contaminated peat the ratio between yttrium and the light lanthanides (Ln) from cerium group: La and Ce is changed. In the Earth's crust Ce:Y-ratio is 2.1 (Greenwood and Ernshaw, 2008). But the background peat is significantly depleted in Ce, and cerium Ce:Y-ratio decreases to 0.33. In the field of oil production, peat is enriched in the light lanthanides from cerium group; as a result the ratio of Ce:Y increases to 1-2. The same applies to lanthanum. If the Earth's crust ratio of La:Y is 1.1, then in the background peat soil is 0.25. In the field of oil extracting, peat is enriched in light lanthanides; a result, the La:Y-ratio increases up to 0.7-1.0. Thus, mainly the lanthanides from cerium group input peat soil and the Ln:Y-ratio is approaching Clarke.

In the course of time, the lanthanides are removed from the oil-contaminated peat. The content of yttrium and lanthanum were reduced by almost half, and cerium - by 1/3. Obviously, the lanthanides migration is determined by the mobile forms of their technogenic compounds.

Table 9. Ni, Br and light lanthanides contents in the ash of polluted peats (ppm)

Key area, point	Depth, cm	TPC %*	Ash, %	Ni	Br	Y	La	Ce
Recently oil-contaminated peat (Ni:Br >1.5)								
№ 09. P.41	0-10	65	7.9	112	69	14	10	13
Ni:Br =2.0	10-30		15.8	63	27	14	16	25
	30-50		44.8	30	6	11	14	23
	50-100		12.8	49	42	13	12	17
№ 09. P. 42	0-10	20	46.8	79	9	44	42	73
Ni:Br =3.5	10-30		24.7	117	19	35	31	57
	30-50		10.9	100	55	34	20	36
	50-100		14.0	123	88	39	24	46
Average						25 ± 5	21 ± 4	36 ± 7
Longstanding oil-contaminated peat (Ni:Br >1.5)								
№ 60. P.16	0-10	60	4.5	200	-	15	21	37
Ni:Br =4.4	10-30		3.9	109	41	16	12	19
waterlogged	30-50		4.1	215	4	14	14	23
	50-100		2.1	56	89	13	6	12
Average						14 ± 0.6	13 ± 3	23 ± 5
Recently saline peat (Ni:Br <0.7)								
№ 73. P. 25	0-10	30	27.0	22	24.0	11	11	16
Ni:Br =0.4	10-30		3.4	35	2.2	14	5	7
№ 60. P.18	0-10		4.9	28	2.8	19	14	17
Ni:Br =0.5	10-30		5.1	24	3.6	16	11	13
	30-50		6.4	31	2.2	10	8	10
Average						7 ± 0.5	13 ± 4	13 ± 5
Longstanding saline peat (Ni:Br <0.7)								
№ 86. P.38	0-10	60	13.5	60	68	8	22	29
Ni:Br=0.04	10-30		3.4	58	20	8	9	12
№ 86. P.39	0-10		12.6	47	147	7	17	21
Ni:Br =0.1	10-30		5.1	24	38	6	5	6
Average						14 ± 1.6	10 ± 1.5	12 ± 2
Background								
№ 13	0-10	100	1.6	17		17	4	6
	10-30		1.4	32		10	2	3
	30-50		2.0	46		10	2	3
Average				32 ± 8		12 ± 2	3 ± 0.6	4 ± 1

*TPC - Total Projective Cover

The lanthanides fate turned out differently in the saline peats. In recent saline peats yttrium content is below the background value, but lanthanum and cerium contents are above background 3-4 times. Over time, La- and Ce-contents are maintained in peat, even Y-content increases. Thus, markedly different behavior of the lanthanides in peat depends on the type of pollution: oil or salts.

Preservation or even the accumulation of Y in saline peat depends probably on the influence of mineral salts. The powerful effect of salts on the clay minerals had been long established in the mineral soils. Destruction of clay minerals, induced usually by organic acids with a high ability to form metal complexes (Vodyanitskii and Rodovneva, 1993), is inhibited in saline soils. In particular, salts inhibit genesis of iron oxides, promoting Fe-conservation in the structure of silicates. Obviously, the minerals affect saline peat. Metal chlorides and sulfates, formed in saline soils, are not able to fulfill the role of complexing agents. Accordingly, in saline soils migration capacity of metal-complexing agents is inhibited. This is what we are seeing in saline peats, where the lanthanides are accumulated because of the impossibility of chlorides and sulfates of these metals to participate in the formation of mobile complexes with organic ligands.

CONCLUSION

The danger of soil contamination by the lanthanides (Ln) and platinum group metals (PGM) increases with the increasing industrialization and urbanization of Russia. The development of electronics, oil chemistry, metallurgy, and the use of modern medical technology increase inflow of Ln-containing waste into soils. Basically, soils are contaminated by Ln and PGM through the aerial way, although hydrogenic contamination of alluvial soils occurs in some places. Increasing the number of post-combustion catalysts fuel vehicles increases air and soil pollution by cancerogenic PGM: rhodium, palladium, platinum. In Russia, dust analysis is widely used in the study of aerial soil contamination by Ln and PGM.

Soils in Russia are contaminated by Ln and PGM near steel plants, thermal power coal-fired plants, as well as in large cities with more car traffic mostly by aerial way.

Some areas of Moscow are contaminated by Ln and PGM. Near Cherepovets town in the Vologda region, soils are contaminated by Ln and PGM, coming from aerial emissions of large steel mill "Severstal." Near Revda in the Ural, soils are contaminated by Ln, coming from aerial emissions of large copper smelter.

Alluvial soils along rivers are hydrogenically contaminated by waters, polluted by untreated industrial wastewater, or soils are contaminated by leachate from dumps, where waste is stockpiled after non-ferrous metals ores enrichment.

Alluvial soils in the Perm city are heavily contaminated by the lanthanides.

In hydrogenically contaminated alluvial soils micropollutants are strongly accumulated in Fe-Mn-nodules, eliminating micropollutants from the geochemical cycle.

Acknowledgments

Authors express gratitude to N. A. Avetov, A. V. Kozheva, E. V. Prokopovich, E. A. Shishkonakova, who provided samples for analysis.

References

Baron, S., Carignan, J. and Ploquin, A. (2006). Dispersion of heavy metals (metalloids) in soils from 800-year-old pollution (Mont-Lozere, France). *Environ. Sci. Technol.* 40, 5319-5326.

Doncheva, A. V., Kazakov, L. K. and Kalutskov, V. N. (1992). Landscape indication of environmental pollution. Moscow. *Ecology*. 240p. [in Russian].

Fiedler, S., Siebe, C., Herre, A., Roth, B., Cram, S. and Stahr K. (2009). Contribution of oil industry activities to environmental loads of heavy metals in the Tabasco lowlands, Mexico. *Water Air Soil Pollut.* 197, 35-47.

Gray, A. L. (1975). Mass-spectrometric analysis of solutions using an atmospheric pressure ion source. *Analyst.* 100, 289-299.

Greenwood, N. N. and Earnshaw, A. (1997). *Chemistry of the Elements.* Elsevier, Ltd. 1342 p.

Ivanov, V. V. (1994-1997). Ecological geochemistry of elements. Volumes 1-6. Moscow. *Nedra-Ecology.* [in Russian].

Kabata-Pendias, A. (2011). *Trace elements in soils and plants.* CRC Press. Roca Raton. 4th edition. 548p.

Kuznetsov, R. A. (1978). *Activation analysis.* Moscow. Atomizdat. P. 219-270. [in Russian].

Ladonin, D. V. (2016). The forms of heavy metal compounds in the technogenic contaminated soils. *Abstract of the thesis of doctor of biological sciences.* Moscow. 42 p. [in Russian].

Migdisov, A. A., Balashov, Yu. A., Sharkov, I. V., Sherstennikov, O. G. and Ronov, A. B. (1994). Prevalence of rare earth elements in the main lithological types of rocks in sedimentary cover of the Russian Plattform. *Geochemistry.* Issue 6, 789-803. [in Russian].

Osovetskii, B. M. and Men'shikova, E. A. (2006). *Natural and technogenic deposits.* Perm State University. Perm. 208p. [in Russian].

Perel'man, A. I. and Kasimov, N. S. (1999). *Landscape geochemistry.* Moscow. Astreya. 768p. [in Russian].

Pikovskii, Yu. N. (1993). Natural and technogenic streams of hydrocarbons in the environment. Lomonosov Moscow State University. Moscow. 207p. [in Russian].

Ravindra, K., Bencs, L. and Van Grieken, R. (2004). Platinum group elements in the environment and their health risk. *The Science of the Total Environment.* 318, 1-43.

Savichev, A. T. and Vodyanitskii, Yu. N. (2009). Determination of barium, lanthanum and cerium contents in soils by the X-Ray radiometric method. *Eur. Soil Sci.* 42, issue 13, 1461-1469.

Savichev, A. T. and Vodyanitskii, Yu. N. (2011). X-ray radiometric determination of the lanthanides contents: praseodymium, neodymium and samarium in soils. *Eur. Soil Sci.* 44, issue 4, 424-432.

Schukova, I. V. (2005). The formation of the chemical composition of a groundwater zone of active water exchange in the city of Perm. Abstract of thesis of PhD of geological and mineralogical sciences. Perm. 23p. [in Russian].

Sharova, O. A. (2009). Environmental aspects of the drilling process and the methods of disposal of drilling fluids. *Geology, geography and global energy*. Issue 4, 29-36. [in Russian].

Sneller, F. E. C., Kalf, D. F., Weltje, L. and Van Wezel A. P. (2000). Maximum permissible concentration and negligible concentrations for rare elements. Report 601501011. RIVN-National Institute of Public Health and the Environment. Biltoven. 66 p.

Solntseva, N. P. (1998). Oil production and geochemistry of natural landscapes. Lomonosov Moscow State University. Moscow. 369p. [in Russian].

Status and protection of the environment in the Perm region in 2003. (2004). Perm State Uviversity. Perm. P. 54-56. [in Russian].

Tyler, G. (2004). Rare earth elements in soil and plant systems - A review. *Plant and Soil*. 267, 191-206.

Vodyanitskii, Yu. N. and Rogovneva, L. V. (1993). Iron oxides in solonetzic chernozems of Stavropol region. *Eur. Soil Sci.* 26, issue 3, 33-42.

Vodyanitskii, Yu. N., Vasil'yev, A. A. and Vlasov, M. N. (2008). Hydrogenic contamination of alluvial soils by heavy metals in Perm. *Eur. Soil Sci.* 41, issue 11, 1399-1408.

Vodyanitskii, Yu. N., Kosareva, N. V. and Savichev, A. T. (2010a). The contents of the lanthanides (Y, La, Ce, Pr, Nd, Sm) and actinides (Th, U) in soils of Khibiny-Lovozero province. Bulletin of Dokuchaev Soil Science Institute. Moscow. Issue 65, 75-86. [in Russian].

Vodyanitskii, Yu. N., Savishev, A. T., Vasil'yev, A. A., Lobanova, E. S., Chaschin, A. A. and Prokopovich, E. V. (2010b). The contents of the heavy alkaline earth (Sr, Ba) and rare earth (Y, La, Ce) metals in technogenic contaminated soils. *Eur. Soil Sci.* 43, issue 7, 879-890.

Vodyanitskii, Yu. N., Plekhanova, I. O., Prokopovich, E. V. and Savichev, A. T. (2011). Soil contamination by emissions of non-ferrous metallurgy. *Eur. Soil Sci.* 44, issue 2, 240-249.

Vorobeichik, E. L., Sadykov, O. F. and Farafontov, M. G. (1994). *Environmental regulation of technogenic pollution of terrestrial ecosystems*. Ekaterinburg. Nauka. 280p. [in Russian].

Voronchikhina, E. A. and Larionova, E. A. (2002). *Basics of landscape chemo-ecology*. Perm State University. Perm. 2002. 146p. [in Russian].

BIOGRAPHICAL SKETCH

Dmitry V. Ladonin

Soil Science Department, Moscow State University, Moscow, Russia

Education: Soil Science Department, Moscow State University, Moscow, Russia

Research and Professional Experience: chemical soil pollution, chemistry of heavy metals in soil, instrumental methods of determination of heavy metals in soil.

Professional Appointments: associate Professor

Publications from the Last 3 Years:

Ladonin, D. V. (2016). Heavy metals compounds in polluted soils. Doctoral thesis. Soil Science Department, Moscow State University: 383 p.

Fedotov, P.S., Ermolin, M.S., Karandashev, V.K. and Ladonin, D.V. (2014). Characterization of size, morphology and elemental composition of nano-, submicron, and micron particles of street dust separated using field-flow fractionation in a rotating coiled column. *Talanta*. 130, 1-7.

Pampura, T.V., Probst, A., Ladonin, D.V. and Demkin V.A. (2013). Lead Content and Isotopic Composition in Submound and Recent Soils of the Volga Upland. *Eurasian Soil Science*. 46, 1059-1075.

Alexandr T. Savichev

Geological Institute, Russian Academy of Sciences;
Dokuchaev Soil Science Institute, Moscow, Russia

Education: Lomonosov Moscow State University, Physical Faculty

Research and Professional Experience: 40 years

Professional Appointments: physical and chemical methods of study of soils and rocks, X-ray fluorescence and X-ray radiometric analyses, rare heavy metals

Publications Last 3 Years:

Vodyanitskii, Yu.N. and Savichev, A.T. (2016). Heavy Metals Contamination of Soils in the Ural Region, Russia. *Contaminated Soils: Properties, Sources and Impacts.* Ed.: Michaela Dunn. Nova Science Publishers, Inc. New York. 121-140. 245p.

Vodyanitskii, Yu.N., Vorobeichik, E.L. and Savichev, A.T. (2016). Heavy Metals as a Biodegradation Inhibitor of the Forest Litter in Contaminated Areas. *Biodegradation: Properties, Analysis and Performance.* Ed.: Jaime Alvarez. Nova Science Publishers, Inc. New York. 201-226. 273p.

Vodyanitskii, Yu.N. and Savichev, A.T. (2015). Improvement of X-Ray Fluorescence and X-Ray Radiometric Methods of Rare Heavy Metal's Diagnostics in Soils. *Advances in Chemistry Research. Volume 25.* Ed.: James C. Taylor. Nova Science Publishers, Inc. New York. 161-180. 205p.

Vodyanitskii, Yu.N., Savichev, A.T. and Kosareva, N.V. (2015). Lanthanides and Actinides in Soils of Khibiny-Lovozero Province. *Advances in Environmental Research.* Volume 41. Ed.: Justin A. Daniels. Nova Science Publishers, Inc. New York. 157-174. 212p.

Vodyanitskii, Yu.N., Manachov, D.V. and Savichev, A.T. (2015). Macro- and microelements including rare earth elements in some soils of the Sakhalin island. *Eurasian Soill Science.* Volume 48, issue 10, 1090-1100.

Vodyanitskii, Yu. N. and Savichev, A. T. (2014). Hydrogenic heavy metals contamination of fluvisols in the Middle Cis-Urals region, Russia. *Water, air and soil pollution.* V. 225, issue 4, 1911-1922.

Ivanovskaya, T.A., Geptner, A.R., Savichev, A.T., Pokrovskii, B.G. and Pokrovskaya, E.V. (2014). Siderite microconcretions in the glauconite-bearing clayey-silty rocks of the Khaipakh Formation (Middle Riphean, Olenek Uplift). *Lithology and Mineral Resources.* Issue 6, 554-582.

Gablina, I.F., Popova, E.A., Sadchikova, T.A., Savichev, A.T., Os'kina, N.S. and Khusid, T.A. (2014). Hydrothermal metasomatic alteration of carbonate bottom sediments in the Ashadze-1 field (13° N Mid-Atlantic Ridge). *Geology of Ore Deposits.* Vol. 56, issue 5, 399-422.

Vodyanitskii, Yu.N. and Savichev, A.T. (2014). Hydrogenic Heavy Metals Pollution of Alluvial Soils in the City of Perm (the Middle Cis-Urals Region, Russia). *Floodplains. Environmental Management, Restoration and ecological Implication.* Ed. E.H. Alcantara. Nova Science Publishers, Inc. New York. 161-180.

Vodyanitskii Yu.N. and Savichev, A.T. (2014). The Distribution of Heavy Metals in Oil-Contaminated Peat Soils in Central West Siberia. *Handbook on Oil Production Research.* Ed. Jacquelyn Ambrosio. Nova Science Publishers, Inc. New York. 219-232.

Yury N. Vodyanitskii
Lomonosov Moscow State University, Moscow, Russia

Education: Moscow Building Engineering Institute

Research and Professional Experience: 40 years

Professional Appointments: Soil Science, Biogeochemistry of Soils

Publications Last 3 Years:

Vodyanitskii, Yu.N. and Savichev, A.T. (2016). Heavy Metals Contamination of Soils in the Ural Region, Russia. *Contaminated Soils: Properties, Sources and Impacts.* Ed.: Michaela Dunn. Nova Science Publishers, Inc. New York. 121-140. 245p.

Vodyanitskii, Yu.N., Vorobeichik, E.L. and Savichev, A.T. (2016). Heavy Metals as a Biodegradation Inhibitor of the Forest Litter in Contaminated Areas. *Biodegradation: Properties, Analysis and Performance.* Ed.: Jaime Alvarez. Nova Science Publishers, Inc. New York. 201-226. 273p.

Vodyanitskii, Yu.N. and Savichev, A.T. (2015). Improvement of X-Ray Fluorescence and X-Ray Radiometric Methods of Rare Heavy Metal's Diagnostics in Soils. *Advances in Chemistry Research.* Volume 25. Ed.: James C. Taylor. Nova Science Publishers, Inc. New York. 161-180. 205p.

Vodyanitskii, Yu.N., Savichev, A.T. and Kosareva N.V. (2015). Lanthanides and Actinides in Soils of Khibiny-Lovozero Province. *Advances in Environmental Research.* Volume 41. Ed.: Justin A. Daniels. Nova Science Publishers, Inc. New York. 157-174. 212p.

Vodyanitskii, Yu.N. (2014). Artificial permeable redox barriers for purification of soil and ground water; A review of publications. *Eurasian Soil Science*. Vol. 47, issue 10, 1058-1068.

Vodyanitskii, Yu.N. and Plekhanova, I.O. (2014). Biogeochemistry of heavy metals in contaminated excessively moistened soils (analytical review). *Eurasian Soil Science.* Vol. 47, issue 3, 153-161.

Vodyanitskii, Yu.N. and Plekhanova, I.O. (2014). Disputable issues in interpreting of the results of chemical extraction of iron compounds from soils. *Eurasian Soil Science.* Vol. 47, issue 6, 697-704.

Vodyanitskii, Yu.N. (2014). Effect of reduced iron on the degradation of chlorinated hydrocarbons in contaminated soil and ground water: a review of publications. *Eurasian Soil Science.* Vol. 47, issue 2, 235-249.

Vodyanitskii, Yu.N. and Savichev, A.T. (2014). Hydrogenic Heavy Metals Contaminations in Middle Cis-Urals Region. *Water, Air, and Soil Pollution.* Vol. 225, issue 4, 1911-1922.

Vodyanitskii, Yu.N. and Savichev, A.T. (2014). Iron compounds in sulfate-carbonate soils on red-colored Cambrian rocks in Southern Angara region. *Eurasian Soil Science.* Vol. 47, issue 5, 553-562.

Vodyanitskii, Yu.N. (2014). Natural and technogenic compounds of heavy metals in soils. *Eurasian Soil Science.* Vol. 47, issue 4, 420-432.

Vodyanitskii, Yu.N., Mineev, V.G. and Shoba, S.A. (2014). Role of zero-valent iron in the degradation of organochlorine substances in groundwater. *Moscow University Soil Science Bulletin.* Vol. 69, issue 4, 175-183.

INDEX

C

F

G

H

I

J

K

L

P

Q

R

S

T

U

V

W

Y

Z